D1598587

DATE DUE

SEP 22 '94			
OCT 11 '94			

HIGHSMITH   # 45220

Anaerobic Digestion:
A Waste Treatment Technology

Critical Reports on Applied Chemistry: Editorial Board

A. R. Burkin	H. J. Cottrell	K. Owen
Chairman	C. A. Finch	K. R. Payne
	C. R. Ganellin, FRS	M. R. Porter
	E. G. Hancock	J. C. Weeks
	A. A. Hodgson	

Critical Reports on Applied Chemistry Volume 31

Anaerobic Digestion: A Waste Treatment Technology

Edited by Andrew Wheatley

Published for SCI
by
ELSEVIER APPLIED SCIENCE
LONDON and NEW YORK

ELSEVIER SCIENCE PUBLISHERS LTD
Crown House, Linton Road, Barking, Essex IG11 8JU, England

Sole Distributor in the USA and Canada
ELSEVIER SCIENCE PUBLISHING CO., INC.
655 Avenue of the Americas, New York, NY 10010, USA

WITH 27 TABLES AND 24 ILLUSTRATIONS

© 1990 SCI

British Library Cataloguing in Publication Data

Anaerobic digestion.
1. Waste materials. Anaerobic digestion
I. Wheatley, Andrew II. Society of Chemical Industry
III. Series
628.354

ISBN 1-85166-526-9

Library of Congress Cataloging-in-Publication Data

Anaerobic digestion: a waste treatment technology/edited by Andrew Wheatley.
  p.   cm.—(Critical reports on applied chemistry; v. 31)
Includes bibliographical references and index.
ISBN 1-85166-526-9
1. Refuse and refuse disposal—Biodegradation. 2. Sewage sludge digestion. 3. Anaerobic bacteria. I. Wheatley, Andrew.
II. Series.
TD796.5.A53 1991
628—dc20   90-37828

No responsibility is assumed by the Publisher for any injury and/or damage to persons or property as a matter of products liability, negligence or otherwise, or from any use or operation of any methods, products, instructions or ideas contained in the material herein.

Special regulations for readers in the USA

This publication has been registered with the Copyright Clearance Centre Inc. (CCC), Salem, Massachusetts. Information can be obtained from the CCC about conditions under which photocopies of parts of this publication may be made in the USA. All other copyright questions, including photocopying outside the USA, should be referred to the publisher.

All rights reserved. No part of this publication may be reproduced, stored in a retrieval system, or transmitted in any form or by any means, electronic, mechanical, photocopying, recording, or otherwise, without the prior written permission of the publisher.

Printed by The Universities Press (Belfast) Ltd.

Preface

Waste treatment by anaerobic digestion has been around for over 100 years. While fossil fuels are plentiful, it has been of little general interest. Information on anaerobic digestion was confined to texts on sewage treatment and ruminant nutrition. This changed in 1974. All forms of renewable energy including those from biomass received a major boost in the wake of the steep rise in the cost of oil.

In 1980 The European Commission predicted that 10–15% of Europe's energy needs could be met by renewables and stimulated a large research and development programme on anaerobic digestion. Virtually every nation that produces technical papers published something on Anaerobic Digestion during the 1980–85 period. The price of oil has now fallen by half compared to 1980, the future price is unpredictable. Other fossil fuels have been reduced in price in competition with oil. This leaves renewable energy in a much weaker financial position.

Circumstances are not the same as they were 10 years ago. Environmental damage resulting from both fossil and nuclear fuel is now an important public issue. Biogas and other renewable energy forms are cleaner than the traditional sources and this will force strategic investment in renewable energy. Next to health care, pollution dominates media coverage of science and technology. Environmental quality is a key part of the status of any area and good health is impossible without a good environment. In northern Europe policital parties whose policies are based on the environment, are a major force.

Waste disposal is also getting more expensive. Implementation of the European Commissions Directives on drinking water, bathing beaches and sludge disposal for example will cost the UK water industry at least £25 billion (billion = 10^9). The UK Government intends that these costs should be incorporated into the charges for water supplied and effluent treated and not be part of general taxation. The charges for water in the UK are the cheapest in Europe apart from Italy and Ireland. Costs in Germany, the most expensive, are 60% higher. Charges will rise at 20% per annum for the next 5 years.

There is therefore a need for further development of efficient waste treatment processes and clean forms of energy generation. Anaerobic digestion fulfills these requirements. This book is intended as a timely re-examination of modern anaerobic technology. The market for anaerobic digestion, the basics of the process and the applications for waste water treatment are reviewed in the five chapters.

The first chapter by Jim Coombs looks at the market for anaerobic digestion. He notes that the early aspirations for anaerobic digestion were not realised. Many of the plants were built hastily with too little investigatory work and failed to reliably achieve the design specification. Most of these pioneer systems were do-it-yourself or collaborative efforts. Since this period the number of contractors offering AD has been consolidated to four or five companies with standard designs all over Europe. Each of these companies has sold more than 20 plants. Jim Coombs finishes his chapter by looking at the prospects for anaerobic digestion of domestic refuse. The gas generated from such sites has previously been an embarrassment or a danger.

Many of the problems associated with anaerobic digestion were caused by misunderstandings about the basic microbiology and biochemistry. In order to control the process properly and be able to take appropriate corrective actions when a problem arises requires a thorough knowledge of the fundamentals. The last five years has seen very great advances in unravelling the complex biochemistry and ecology of anaerobic digestion. This process is still going on. David Archer and Brian Kirsop in the second chapter describe the microbiology of the process. It has emerged that the synergism between the different organisms in anaerobic digestion is vitally important. Disruption to any group or the ecological balance can cause major problems. The biochemistry and conditions under which these bacteria live are unusual and temporary additions of extra nutrients or adjuncts can help to overcome many short term difficulties. Much is yet to be learnt about these organisms and additional useful industrial applications are confidently predicted.

Peter Hobson in his chapter looks at the current role of digestion for the treatment of farm wastes. All of the most recent Water Authority annual reports and other Government statistics point to the farming community as a major source of pollution of rivers and ground water. His conclusions leave the question open, improved simpler anaerobic technology is available but it is still expensive and more time consuming than dumping for most family-run farms. The answer is in part already provided by the changing role of farming but also relies on better persuasion of the farming community. Further rationalisation of the farm sector into larger units will also contribute to solving the problem. Parallels can be drawn with small digesters in rural communities. Ignorance of the process basics, bad mixing, heating and poor materials of construction have led to many failures. This leads us directly to Gerald Noone's chapter. These were precisely the same problems he faced when inheriting the sludge disposal questions at the Severn–Trent Water Authority. Massive, expensive, inflexible concrete digesters which were impossible to mix and heat were the norm. The theme of Gerald Noone's paper is the application of standard process engineering techniques. He started from the fundamentals of the process and rebuilt anaerobic digesters with 'off the shelf' prefabricated equipment. The result was a radical change in digester design and a cheaper, more flexible and reliable process for the treatment of sewage sludge. Gerald Noone has recently been awarded an MBE for his contribution to waste management.

Finally we consider the anaerobic digestion of industrial waste. Few industrialists need much more data than the economics of the process to convince them of the benefit. In the UK however applications have been held back by a lack of full scale experience. The UK waste treatment industry by comparison with Europe is smaller, fragmented and less able to support development costs. The market for waste treatment will shortly receive a boost in the form of a steep rise in water costs associated with privatisation and harmonisation. Industry will be using a lot less water in the future. More precise control and integrated processes will reduce wastage. The smaller volumes of more concentrated waste will be ideal for anaerobic digestion. For some industry with very strong waste, distilling and fermentation for example, there are few alternatives.

All the contributions to this book therefore point to a greater use of anaerobic digestion in the future. We hope that these reviews will be a useful update on the technology for Water Authorities, Industrialists, Agricultural Engineers and Academics with an interest in waste treatment.

ANDREW WHEATLEY

Contents

Preface v

List of contributors x

1 **The present and future of anaerobic digestion** 1
   J. Coombs

2 **The microbiology and control of anaerobic digestion** 43
   D. B. Archer and B. H. Kirsop

3 **The treatment of agricultural wastes** 93
   P. N. Hobson

4 **The treatment of domestic wastes** 139
   G. P. Noone

5 **Anaerobic digestion: industrial waste treatment** 171
   A. D. Wheatley

Index 225

List of contributors

David B. Archer	AFRC Institute of Food Research, Norwich Laboratory, Colney Lane, Norwich NR4 7UA, UK
James Coombs	CPL Scientific Ltd, Science House, Winchcombe Road, Newbury, Berks RG14 5QX, UK
Brian H. Kirsop*	AFRC Institute of Food Research, Norwich Laboratory, Colney Lane, Norwich NR4 7UA, UK
Peter N. Hobson	Department of Molecular and Cell Biology, Marischal College, University of Aberdeen, Aberdeen AB9 1AS, UK
Gerald P. Noone	Severn–Trent Water Authority, Abelson House, 2297 Coventry Road, Sheldon, Birmingham B26 3PU, UK
Andrew D. Wheatley	The Biotechnology Centre, Cranfield Institute of Technology, Bedford MK43 0AL, UK

*Present address: Biostrategy Associates, 10 Waterside, Ely, Cambridge CB7 4AZ, UK.

1 The present and future of anaerobic digestion

J. Coombs

CPL Scientific Ltd, Newbury, UK

1.1	Introduction	1
1.2	Products and uses	5
1.3	Design considerations and AD markets	9
1.4	Rural digesters	13
1.5	Farm digesters	15
1.6	Sewage sludge digestion	21
1.7	Industrial effluent digestion	24
1.8	Municipal solid waste digestion	28
1.9	Landfill gas	31
1.10	Biogas energy systems	34
1.11	Conclusions	35
1.12	References	36

1.1 Introduction

During the process of anaerobic digestion (AD) organic raw materials are converted to biogas, a mixture of carbon dioxide and methane with traces of other constituents, by a consortium of bacteria which are sensitive to or completely inhibited by oxygen. Using AD it is possible to convert rather intractable plant residues, agricultural wastes, manures, effluents from the food and beverage, paper and pulping and some chemical industry wastes into useful by-products. The organic fraction of municipal solid waste (MSW) or purpose grown crops including conventional species and aquatic plants such as water weeds or algae may also be digested to give a potentially useful fuel which may be variously used to provide heat, electrical power or transport.

As detailed in Chapter 2 the bioconversion of organic materials to methane is accomplished by chemoheterotrophic, non-methanogenic and methanogenic bacteria, with larger, polymeric compounds first hydrolysed to free sugars, alcohols, volatile fatty acids, hydrogen and carbon dioxide. This mixture is oxidized to acetic acid, carbon dioxide and hydrogen which are then converted to methane. Lignin is not degraded by most AD systems; the rate of cellulose breakdown is slow (weeks), hemicellulose and protein somewhat faster (days) and small molecules such as sugars, fatty acids and alcohols fast (hours). Variations in feed composition or temperature resulting in 'shock' can cause an imbalance in microbial activity resulting in changes in pH, gas composition and efficiency of COD removal. Stability is a major concern in commercial systems and may be controlled by addition of alkali as well as control of feedstock composition and feeding rate.

The quest for new forms of renewable energy in both developing and developed countries stimulated worldwide research, development and demonstration interest in AD in the 1970s. However, a decrease in the real price of fossil fuels, failure of many digesters and an increasing concern about environmental pollution resulted in a shift in emphasis from energy to using AD as a means of treating both effluents and solid wastes. During the same period increases in the amount of municipal solid wastes (MSW) disposed of to landfill resulted in deeper and larger accumulations. This encouraged AD and resulted in problems associated with methane leaking from the landfill site. Attempts to control the migration of such landfill gas (LFG) resulted in its use as a source of energy.

Historically, AD had been associated with the water industries for the consolidation of sewage sludge with the resulting biogas used to heat the sludge digesters and generate electricity for in-house use. Knowledge from sewage digestion, including reactor design, microbiology, and gas use formed the base from which a wide variety of digester designs evolved in the 1970s. The objective was to produce biogas as an energy source during the treatment of industrial, commercial and agricultural wastes or effluents. This development was supported by considerable basic research resulting in a better understanding of the microbiology and biochemistry of the process (see Chapter 2) as well as the development of many different types of bioreactors (see Chapter 5). At the same time the technology required for use of the gas was also developed. Gas clean-up processes were improved, burners and boilers were modified and engines, both internal combustion or turbine-driven electricity generators, were made more reliable. Compression systems to enable the gas to be used as a vehicle fuel were also developed (Constant et al., 1989).

In theory AD offers significant advantages in energy terms over aerobic waste treatments which require an energy input and produce little of value. However, a comprehensive survey (Demuynck et al., 1984) covering the development of AD on farms and as a means of treating industrial effluent within the EC during this period has shown that many projects, schemes or national programmes have not been as successful as anticipated.

In Europe for instance, early estimates (Anon., 1983) suggested an energy resource in wet wastes, effluents and manures equivalent to between 5–10% of current demand with a realizable potential of over 11·5 mtoe (million tonnes oil equivalent) from animal wastes, 12·5 mtoe from crop wastes, and 8·9 mtoe from MSW; a total of over 30 mtoe or 3% of European energy consumption in 1985. However, in reality energy contribution from the first phase of AD construction in the EC was insignificant (Demuynck et al., 1984). A total of 95 000 m^3 of digester volume had been installed on farms at this time as well as 173 000 m^3 in industrial plant. On an optimistic basis of 1 m^3 biogas per m^3 reactor volume and a 30% effective use this is equivalent to around 33 000 toepa (tonnes oil equivalent per annum), with a value of only £3–4 million, much less than had been spent on research and development programmes.

The situation in the UK has been similar. Optimistic claims based on resource surveys in the late 1970s suggested that biogas could be produced in significant amounts. Farm wastes were expected to generate 2·51 mtcepa (million tonnes coal equivalent per annum), domestic wastes 3·34 mtcepa and crop wastes 4·62 mtcepa. It was further anticipated that energy crops grown on 7 million hectares could generate a further 18–26 mtcepa. Subsequent estimates were much more conservative with the greatest contribution in the short term expected from landfill gas (0·33 mtcepa by 1995). The production of landfill gas and the anaerobic digestion of sewerage and industrial effluents was regarded as economically viable at 1985 fuel prices with an energy potential of 0·8 mtcepa. It is now thought that the use of animal and crop wastes could contribute only 1·6 mtcepa by 2025 (Anon., 1987a). In excess of £10 million has been spent on research and development of anaerobic digestion by bodies such as the Department of Energy, Ministry of Agriculture, Fisheries and Food, Department of Trade and Industry and the Science and Engineering Research Council. This has resulted in the construction of around 20 farm digesters and nine industrial plants.

The viability of AD can be judged from a commercial viewpoint in terms of the present market size, the growth rate and future expectations as well as considerations of the economic strength of companies involved in AD. In evaluating the present situation it must be realized that the industries based on modern AD technology and LFG extraction are young. The EC survey (Demuynck et al., 1984) indicated that significant building of farm digesters started in 1979–80,

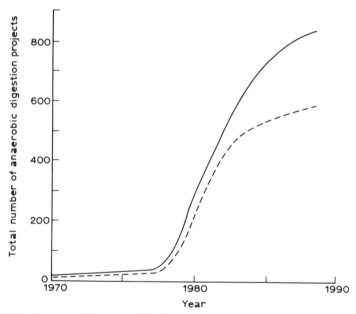

Fig. 1.1 Growth of the anaerobic digestion industry within Europe (Figures for member states of the EC, not including Spain or Portugal, together with Switzerland; based on Demuynck et al., 1984 and Pauss et al., 1987). Cumulative number of projects of all types ———; cumulative number of farm-based projects - - - - -.

industrial digesters a year or so later, with landfill projects starting to be commissioned in 1982–83. In other words, the industry on any scale is less than 10 years old (Fig. 1.1).

Similar variations in the performance of AD in other countries, both developed and Third World, support a conclusion that there are still technical problems and few total successes leaving governments, funding agencies and potential industrial users unsure of the true value of AD. Hence, at the start of the 1990s AD stands at a threshold with a need to build on the knowledge gained in the 1970s and 1980s (Hall and Hobson, 1988) in order to realize the full potential of this technology.

Significant changes in AD markets are already apparent (Fig. 1.1). Preliminary results updating the EC survey (Pauss *et al.*, 1987*a*) show that in the farm sector only about 40% of plant previously reported appears still to be working, few new digesters have been built and a substantial decrease has occurred in the number of farm digester constructors still in business. In contrast the number of digesters treating industrial wastes in the EC continues to increase by 20–30 a year with the number of designers and constructors of industrial systems remaining fairly constant.

1.2 Products and uses

The value of an AD system can be evaluated either in terms of the cost of disposal of organic wastes, or in terms of the value of products. The main product in many cases is biogas. Hence, the success of many projects depends on the net fuel value of the gas produced in a given time, which in turn reflects composition and volume as well as the value of other fuels for which it can be substituted on the one hand and the capital investment and annual running costs on the other.

Table 1.1 summarizes average gas composition and calorific values for biogas from various sources. As indicated the properties and the volumes of gas produced vary widely depending on microbiological, engineering, temperature and feedstock related factors. Digestion of most common feeds will give equal volumes of methane and carbon dioxide although fats give a gas containing 72% methane. The biogas will in general be enriched in methane since carbon dioxide has a greater solubility in water than methane and will tend to form bicarbonates at pH 7 typical of many AD systems. The average

Table 1.1 Constituents of biogas derived from different sources

Constituent gas	Agricultural wastes	Sewage sludge	Industrial wastes	Landfill gas	Comments
Methane	50–80%	50–80%	50–70%	45–65%	Fuel
Carbon dioxide	30–50% (40)[a]	20–50% (40)	30–50% (40)	34–55% (45)	Diluent, acid, asphyxiant
Water	Saturated	Saturated	Saturated	Saturated	Corrosive, reduces CV
Hydrogen	0–2%	0–5%	0–2%	0–1%	Fuel
Hydrogen sulphide	100–7 000 ppm (500 ppm)	0–1% (100 ppm)	0–8% (<0·5%)	0·5–100 ppm (30% recorded)	Corrosive, toxic, odour
Ammonia	Tr	Tr	Tr	Tr	Corrosive
Carbon monoxide	0–1%	0–1%	0–1%	Tr	Toxic
Nitrogen	0–1%	0–3%	0–1%	0–20%	Diluent
Oxygen	0–1%	0–1% (0·2%)	0–1%	0–5%	Corrosive
Trace organics	Tr	Tr	Tr	5 ppm (terpenes, esters, hydrocarbons)	Odour, corrosive

[a] Average or common value.
Calorific value (CV) 17–34 MJ/m^3 depending on methane content.
Tr—trace amounts.

calorific value of biogas will thus be around 20 MJ/m^3. Methane content may be reduced as a result of decreased activity of the methanogenic bacteria as a result of acid accumulation (due to overloading, temperature or other shock effect) or as a result of poisoning by contaminants such as biocides, phenolics or heavy metals.

Both organic sulphur (present in amino acids for instance) and inorganic sulphur (gypsum present in building wastes or generated from neutralization of sulphuric acid added to fermentation feedstocks such as molasses to remove excess calcium) may be reduced to hydrogen sulphide. This gas is extremely toxic as a respiratory poison and highly reactive with metals such as copper and iron, causing corrosion. Such corrosion may be a major problem in shortening the life of some types of internal combustion engines. It may also be toxic to the methanogenic bacteria (Chapter 2). Conversely, the high reactivity of this gas with metals in landfill sites accounts for the generally low content of hydrogen sulphide in LFG. Organic nitrogen is reduced to ammonia most of which stays in solution reacting with hydroxyl ions generated during protein deamination. These alkaline radicals also increase bicarbonate formation resulting in a gas of higher methane content and an effluent of higher nitrogen content (and hence fertilizer value) from feeds, such as manures, high in protein.

Volumes of gas produced on both a substrate (m^3/kg COD) and a volumetric (m^3/m^3 digester volume) basis differ with the type of waste, concentration of volatile solids (VS), loading rate, retention (detention) time and reactor design. In general production rates of between 0·3–0·5 m^3/kg VS might be expected with between 1–5 m^3/m^3 digester volume although, as detailed below, large variations are found in operational digesters.

Condensation is often a problem since the water-saturated gas from the digester is generally warmer than the pipework into which it passes. Water may also be carried over in the gas as an aerosol or even as a foam in systems which are vigorously stirred. In smaller systems water traps and drain points are essential, although many problems have arisen from the use of wrong size pipes or poorly orientated pipe runs. In larger systems gas chilling to remove water may be essential. Hydrogen, an intermediate in anaerobic metabolism, may be expected in LFG from young sites. In fabricated digesters it may indicate imbalance or inhibition of methanogenic activity and can be used to monitor digester performance (see Chapter 2). Traces of carbon monoxide may be produced by some anaerobes.

Anaerobic bacteria are highly versatile and can degrade many compounds, including some xenobiotics, generating a wide variety of products. These compounds are found in trace amounts in LFG in particular due to the wide variety of materials present in industrial waste which may give rise to aromatic compounds, mercaptans and/or halogenated compounds including higher hydrocarbons, terpenes and esters (Anon., 1986a). The presence of nitrogen and/or oxygen in biogas indicates air leakage which in fabricated digesters can represent a serious hazard due to the risk of gas explosions. Oxygen is less likely to be seen in LFG since air which is drawn into the landfill is usually consumed in aerobic decomposition resulting in residual nitrogen, high levels of which are an indication of excessive gas removal.

The EC survey (Demuynck et al., 1984) found that the major on-farm uses for biogas were for heat and/or electricity generation with combustion in boilers the main industrial use. Economics improve as the percentage gas use increases and as the value of fuel substituted or power generated increases.

In many cases however, there is an imbalance between production and in-house demand, resulting in need for back-up energy supplies in case of shortfall or identification of an alternative use for surpluses. Domestic uses of the gas are for cooking where the demand may be between 2–6 m^3 per hour or for space heating using a water boiler. Space heating is also a common use of the gas in agriculture and industry (giving around 5·3 kilowatt per m^3 gas consumed). Other industrial process uses include the generation of high pressure steam as well as brick-making, lime kilning or cement production. Such activities may be associated with the production of large amounts of LFG which may be generated locally by tipping domestic refuse in the worked-out gravel, clay or cement pits. LFG may also be piped to nearby facilities and used in industrial processes such as iron smelting, metal-working, glass production and pottery firing.

Spark ignition or dual fuelled engines are easily adapted to run on biogas and may be used to produce power for pumps or tools or to generate electricity for on-site use or export to the grid. The feasibility and profitability of any particular option is site-specific, but generally increases where large amounts of gas are available on a reliable basis for extended periods. For electricity generation the cost of equipment and transmission lines needed to facilitate input to the grid and the tariffs available in some countries make this option unattractive to the small producer. Frequently however, the control of LFG migration is essential and if no industrial use can be found, then electricity generation with supply to the grid may be the most attractive option. The economic benefit of using biogas as a vehicle fuel depends on availability and price of conventional liquid fuels. Both diesel and spark ignition engines may be adapted in the same way as for CNG (compressed natural gas) or LPG (liquefied petroleum gas) to run on methane, but 1 m^3 of gas is equivalent to approximately 0·8 litres of petrol powering an average car about 8 km, or a lorry 2 km. To achieve a reasonable total mileage per fuel charge the gas must be compressed, after purification, with a need to use bulky fuel tanks, or liquefied at much higher costs. In either case the vehicle will probably be restricted to a range of 80–120 km from the biogas source. The degree of purification required for pipeline distribution, either to supplement other gas supplies or to fuel dedicated equipment working only on biogas varies from chilling to remove water and trace organics to production of methane at over 98% purity for addition to national natural gas supply systems.

In addition to gas AD may generate other products which can be valued or sold as well as having a number of less tangible benefits such as pathogen control. The most widely used by-product of AD is the effluent which, depending on residual solids, nitrogen content or water purity, may be used as fertilizer, soil conditioner or for irrigation. The microbial biomass has been used as a single cell protein supplement in animal feed (Mowat *et al.*, 1986). However, problems will arise as a result of poor feed value or digestibility as well as from contamination by heavy metals, bacteria, parasites and other pathogens. In the case of animal wastes, antibiotics or synthetic growth hormones may also be present. Hence, such use is unlikely.

Anaerobic digestion systems have an important role in control of odours from farm wastes and sewage sludges (Nielsen *et al.*, 1986) whilst the combustion of LFG, whether used for energy or flared, is an essential means of controlling odour from landfill sites (Anon., 1986*a*). Mesophilic digestion will kill some pathogens such as *Salmonella* (Bruce, 1987) and viruses during retention times of 20–30 days (Havelaar, 1985), completely inactivating enterovirus ECBO-LCR-4 and Reo (Reovirus Type 1). However, others such as bovine parvovirus are found to survive even thermophilic temperatures. Similar studies show that the infectivity of tapeworm (*Taenia saginata*) eggs is greatly reduced, but eggs of *Ascaris suum* survive, again even at thermophilic temperatures.

1.3 Design considerations and AD markets

The rate of degradation of organic material reflects the interaction of the amount of active microbial biomass per unit volume and organic material available to the biomass as substrate. A major design objective is to achieve a high biomass content within the reactor in order to achieve both a high gas production and a high COD reduction per unit of digester volume from a high organic loading rate. For a given throughput the digester size is determined from the solids content of the waste and retention time. Wastes high in insoluble material such as paper, straw and other lignocellulosic materials may require treatment for days (or even years as in landfill sites) whereas up to 95% reduction of the organic material (BOD) at loading rates of over 20 kg per day per cubic metre of reactor may be achieved with process effluents where the waste is soluble.

Most commercial digesters are run in the mesophilic range (30–40°C) requiring some heating although some systems run at ambient temperatures in the psychrophilic range (around 20°C) and a few thermophilic systems are operating at over 50°C. In the laboratory, however, the upper temperature limit of AD has recently been raised to 97°C and evidence for bacterial methanogenesis at over 100°C comes from studies of extremeophiles from the submarine volcanic environments of the East Pacific Ocean (Deming, 1987). These high temperature systems have not yet been exploited commercially. With wastes of high insoluble solids content unstirred batch reactors, stirred tanks or plug-flow systems may be used. However, for dilute effluents there is a need for processes with long solids retention times (SRT) and short hydraulic retention times (HRT). High SRT results in process stability and minimal sludge production, whereas a low HRT reduces reactor volume and hence capital costs. The slow growth rate of the methanogenic bacteria requires an efficient means of maintaining a high level of viable biomass in the reactor. This can be done by separating the cells by centrifugation, sedimentation, flotation or filtration and recycling them back to the reactor as in the Contact Process. Alternatively the conditions during start-up may be manipulated in such a way that the organisms form granular beads as in upflow anaerobic sludge blanket (UASB) processes, or on particulate supports as used in fluidized or expanded bed reactors. Other immobilized cell reactors include anaerobic filters in which the cells grow on a bed of random (often plastic) support material of high surface area (Chapter 5).

Problems of variation in the rates of acid production and methane generation have been tackled using two-stage systems. The first stage may consist of a reactor packed with solid feedstock, from which leachate is fed to a second anaerobic filter. If the objective is to produce water to meet environmental discharge levels it may be necessary to combine the anaerobic process with an aerobic polishing system. Various types of digester tend to be matched to specific feeds. Many Third World rural digesters are of the unstirred batch type, most farm digesters are stirred tank or plug-flow reactors, most sewage digesters are stirred tanks whilst many of the more complicated (sometimes termed second generation digesters) are associated with treatment of industrial effluents.

The economics of anaerobic digestion are difficult to quantify; in general they need to be carried out on a site-specific or project basis.

In most developed economies a simple balance sheet of capital cost plus running costs against the revenue from the sale of gas (or equivalent value of gas in terms of cost of oil or other fuel saved) is not relevant. The other incentives of tax credits, pollution legislation, odour control or waste disposal are more important. The level and availability of subsidies in one form or another for AD plant has varied both with time and from state to state in the EC (Demuynck et al., 1984). However, the impact of such subsidies is variable. Taking the results of the EC survey as an example, Switzerland (108), Germany (75), France (74) and Italy (63) accounted for the largest number of farm digesters. In Switzerland the digesters were less than 5% part-funded, whilst in France there were at least seven different sources of support and in Italy owners could receive from 30% to 70% subsidies. In the UK two or three of the digesters had received support through EC and/or government demonstration funding.

Methods of assessing the performance of a proposed digester system have been suggested by Chesshire (1986), using the effectiveness of the system expressed in terms of either solids reduction or gas production per tonne of solids loaded. On this basis if the measured specific gas production (Y m^3 of gas per tonne of dry matter) is close to the theoretical then the digester is working well. A further index of performance can be based on solids throughput. The detention factor (D) based on digester retention time (R in days) and therefore digester volume, divided by the proportion of dry matter (S).

$$D = R/S$$

These two efficiency parameters may be combined to give a single measure of the feasibility (F) of a digester system in terms of gas production per capital cost: feasibility (F) = gas production divided by the digester volume. Hence,

$$F = Y/D$$

The higher the value of F the lower the capital cost of the system. Examples are given by Chesshire (1986) showing that the value of F for a sewage digester can vary from 0·3 ($R = 30$ days, $S = 0·03$) to 2·5 ($R = 10$ days, $S = 0·08$). For farm systems it is suggested that with a value of F of less than one the system will not be viable as an energy producer.

Some economic information was obtained during the first EC survey

(Demuynck et al., 1984). This was limited by the fact that many projects had not yet stabilized or had not been properly monitored. However, data were obtained for 32 farm digesters which showed that only six had a pay-back period of less than 6 years and over half had a pay-back of over 10 years. The main reasons for poor economic value were identified as poor performance, poor gas use, high maintenance costs and an unduly large need to use gas to heat the digester. It was concluded that to be economic the minimum requirement is the production of 1 m^3 biogas per m^3 reactor per day. This level was seldom realized in farm digesters, although much higher volumetric gas production rates are observed with AD plant treating industrial effluents, as discussed below. An economic capital investment of less than £300–400 per m^3 digester volume was also seldom met in the farm sector.

Techniques for economic analysis of specific projects have been suggested (Biljetina, 1987; Durand et al., 1988). However, detailed analysis may be hampered by the lack of information on total capital costs of digester systems. Many manufacturers are wary of publishing detailed cost information and in some cases digesters are over-designed or over-sized, since the objective of the manufacturer will be to meet process guarantees and possibly maximize plant and equipment sales. The small number of manufacturers in many countries results in little 'price competitiveness' whilst at the same time total sales are insufficient for the normal factors of cost reduction associated with multiple units. Again the site-specific nature of projects as mentioned above may be such that very little replication takes place anyway. These problems have been stressed by Biljetina (1987) who points out that

> it should be recognized that the important design and operating data are not readily available to the scientific community. Digester and operating information is frequently proprietary and protected by secrecy agreements. Consequently development efforts are fragmented and no concensus has yet emerged on design, operating conditions and appropriate end uses.

There were insufficient operating data from a bewildering array of existing and planned systems.

In fact the problems are frequently worse than this, due in part to the conflicting claims and commercial propaganda put out by some manufacturers who overstate the reliability and performance of their systems. This has resulted in independent monitoring studies of both government funded and independently constructed digesters. Companies have also resorted to marketing strategies in which the digester is installed on a contract basis, with the manufacturer managing the plant and being paid on the basis of results rather than selling the system on a conventional turn-key basis.

Of the early farm and industrial AD systems one-third were do-it-yourself (DIY) systems and the rest either turn-key plant or collaborative efforts between a constructor and the client with few companies having managed to build sufficient plant to establish a track record. The situation as reported in Italy at the end of 1983 (Anon., 1984) was typical of most countries. About 40 independent organizations had been involved in the construction of 64 digesters with a total volume of 44 600 m^3 constructed plus another 22 (20 000 m^3 total capacity) under construction. Of these only one organization had reached double figures (11), 20 had built only one digester and seven had built only two. The large number of digesters built serves to emphasize the number of companies or other concerns which have only built one digester, and hence the problem in establishing either standard technology or a track record.

1.4 Rural digesters

Rural digesters are in general small with active volumes of between 1–10 m^3, fed in a batch manner with human wastes, animal manures and crop residues. Progress in transfer and diffusion of rural AD technology has been summarized by El-Halwagi (1986). Most rural digesters, typified by Chinese or Indian designs, located in developing countries provide small amounts of gas used for domestic heating and lighting. The Chinese style digester is generally sunken, built from bricks or concrete with the gas constrained within the fixed digester volume. In contrast the Indian style (Gobar) digester, constructed of iron, is generally of the floating gasholder type. Another small-volume simple type of digester is the Taiwan plastic digester, based on a long tubular bag, working in a semi-plug-flow mode (Stuckey, 1986). A wide range of other types of small AD plant have also been built (Edelmann, 1986; Ward, 1986), often with locally available materials such as wood, cement, plastic, metal, fibre glass, brick or compacted mud.

Rural digesters have been built in large numbers during the last 10–20 years. Biljetina (1987) quotes figures of 5–6 million in China, 1 million in India, 30 000 in South Korea, 2300 in Brazil and 1200 in Nepal. However, Smil (1986) suggests that in China, even during the peak period of construction, less than half were working, and of these not many could be used throughout the year to cook rice three times a day. Once the end of the rigid political regime permitted more freedom, many of the digesters were abandoned. Assuming some 4 million, with an average volume of 8 m^3, generating 0·2 m^3/m^3 digester the total gas production at around 1·25 mtcepa would be less than 0·5% of China's rural energy use.

Much of the emphasis on rural digestion has occurred in Asia, reflecting in part the growing fuelwood deficiency in the Far East. Moulik et al. (1986) report that by 1983 over 0·28 million digesters had been built in India, with a crash programme aimed at building 6 million units at a rate of 0·15 million pa starting in 1984–5. In Nepal, 1600 family-size units and 24 community plants have been built (Gorkhali, 1986) whilst Thailand has over 5000 family-sized units and around 10 community projects (Chantavorapap, 1986). Orcullo (1986), suggests there are around 600 digesters, generating a total of 26 000 m^3 per day of biogas in the Philippines but puts the potential at over 1·28 billion (10^8) m^3 per year. Economics are reasonable for the locally built systems with pay-back periods of from 2–4 years for large systems and of 3–5 years for domestic models, with the industry supported by at least six commercial companies. However, development of the market is slow with projections suggesting that numbers may not pass 1000 until 1992, a long way off realization of the suggested potential.

Outside the Far East, Caceres and Chiliquinga (1986) have described a total of 3950 digesters in 21 Latin American countries, of which over 3000 are in Brazil. Almost all are less than 100 m^3, many of Indian or Chinese design. In Egypt a national AD programme, started in 1981, aims to increase the efficiency of rural energy production from the present level of less than 10% for direct combustion of plant residues and dried manure to over 30% (Alaa El-Din and El-Shumi, 1986). By 1985 only 40 units, again based on Chinese, Indian or Taiwanese designs varying from 6 to 200 m^3, had been constructed. Although costs vary greatly from country to country, depending on design and materials used, Pluschke (1986) has shown a significant scale factor even in small digesters with total costs ranging from around £320 per m³ digester volume for small digesters of 3 m³ to around £30 for 12 m³ systems of Indian design. At these costs AD can offer rural communities a reliable and convenient source of energy for domestic, agricultural and light industrial use with both social and economic benefits for individuals and communities.

Many bodies including Latin American organizations (OLADE and ICAITI), French (GERES), German (GRZ/GATE) and US aid agencies and the Rural Energy Network of the FAO have promoted rural AD (El-Halwagi, 1986). These are in addition to national programmes and projects in Botswana, Burkino Faso, Ivory Coast, Kenya, Lebanon, Mali, Morocco, Sudan, Tunisia, Uganda, Sri Lanka, the Caribbean and elsewhere. However, a combination of poor performance, problems of maintenance, labour requirements and high costs when compared with fuelwood plantations has led to doubts in some sectors of the value of promoting rural AD on a national scale.

1.5 Farm digesters

Most farm digesters treat manures from cattle, pigs or fowl with the size of digester reflecting local practice in terms of herd size or fowl number with installations typically in the range of 30–1000 m³ designed to produce 1 m³ per m³ digester volume/day, although some larger systems have been built in the US and elsewhere with wastes from several farms or animal units being brought to a common AD facility. A very wide range of structures have been used on farms ranging from simple batch-fed, unstirred, membrane covered slurry pits, metal or fibreglass tanks which include a gas-holding space above the liquid effluent, through to large plug-flow or conventional completely stirred tank reactors (CSTRs) with discrete gasholders. Many systems are based on grain silos or similar tanks with mechanical stirring or liquid and/or gas recirculation. The mesophilic temperature range is most common, with digesters heated by combustion of part of the gas or indirectly using engine cooling water where the plant includes electricity generation. A few filter systems have been used to treat liquid following solids separation.

From an operational viewpoint heating can be a problem in temperate countries in winter where manure is allowed to freeze, is washed off with excessive amounts of cold water or stored for long periods prior to digestion. The high level of hydrogen sulphide often found in biogas generated from manures has caused considerable problems in electricity generation through damage to IC (internal combustion) engines. More discouraging is the number of design or construction problems identified in the 1983 EC survey which included gas leaks, poor feed quality, poor gas production, poor mixing and poor engine performance (Table 1.2).

Conclusions (Friman, 1986) on the value of farm AD systems in the UK were equally negative with fewer than 20 digesters (of which only eight were larger than 50 m^3) operational. Results of monitoring studies carried out by MAFF on five of these showed that only one consistently gave over 1 m^3 biogas/m^3 reactor. The poor performance (0·5–0·8 m^3/m^3 digester) was associated with over-dilution of pig slurries although the specific gas yields (0·28–0·34 m^3/kg total solids (TS) fed) were reasonable suggesting that a TS of at least 60 kg/m^3 is required for a viable system. A plug-flow cattle manure digester also performed less well than anticipated, due to poor mixing and heating of incoming cold undiluted cattle manure with a TS of over 100 kg/m^3, suggesting that in this case dilution might have improved the situation. Better results were achieved in an experimental full-scale plant using two 125 m^3 mixed digesters at a 400 cow unit to compare gas production from separated and unseparated slurry (7% solids) (Pain et al., 1984). Steady state gas yields (55% methane, 22 MJ/m^3) averaged 0·2 m^3/kg TS for separated slurry at 20 days retention time (RT) compared with 0·28 and 0·25 m^3/kg TS for separated slurry at 20 and 15 days RT respectively. Although separation reduced gas production by 30% the overall effect was to reduce the VS and COD of the liquid effluent by 70% and 65% respectively.

Studies on five digesters in Italy have been reported by Bonazzi et al. (1984). The digesters were operated at a HRT of between 28 and 40 days, using pig, cattle and mixed wastes. Gas production of around 0·28–0·5 m^3 kg VS/day were recorded with poor volumetric yields (0·36–0·62 m^3/m^3 digester/day) again associated with poor slurry quality. In Spain, of 28 farm plants built prior to 1984 fewer than 40% were found to be operating normally (Garcia Buendia, 1987).

In Austria around 30 AD plants had been built on farms by 1985 of which eight, gas-stirred digesters with sludge recycle, were built by the same company. Analysis of performance and costs of ten plants (Pernkopf, 1986) again proved disappointing with an annual cost (depreciated over 10 or 15 years with a 10% interest rate) of between ASch. 35 000–90 000 and generated energy worth only ASch. 12 000–30 000.

Table 1.2 Problems found in anaerobic digestion systems; based on Demuynck et al. (1984). (Number of cases out of a total of 96 answers analysed)

	Major problem	Minor problem
Feedstocks		
Settlement on storage	8	8
Rainwater entry	8	4
Feed variation in factories	2	14
Scum formation	6	4
Methane production/volatile loss	4	16
Loading		
Pipe blockage	6	15
Feed heating problems	8	3
Alkali precipitates in lines	5	3
Problems with heat exchangers	2	12
Digester engineering		
Gas leaks	13	14
Poor mixing	11	8
Poor heat transfer	6	5
Electrical failures	5	2
Problems with safety valves	2	7
Corrosion	1	9
Problems with insulation	2	2
Digester operation		
Settlement of solids	6	9
Scum formation	7	5
Poor gas separation	2	6
Outflow blockage	2	0
Start-up problems	2	0
Gas collection, storage and use		
Water blockage of lines	1	17
Gas pumping problems	6	5
Leaking gas holders	3	5
Problems with burners	1	12
Problems with engines	12	3
Problems with waste heat recovery	9	9
Chemical problems		
Hydrogen sulphide	3	16
Buffering	4	2
Ammonia toxicity	2	4
Antibiotic toxicity	2	1
Monitoring problems		
Gas metering problems	19	10
Temperature probe problems	3	8
pH probe problems	3	7

For Germany Perwanger (1986) reported a detailed analysis of 25 farm digesters between 30 m^3 and 1500 m^3, mainly dealing with cattle manure. Productivity was much lower (0·7 m^3 biogas/m^3 reactor volume and 0·29 m^3/kg VS) than promised by the constructors with a low cost DIY system of 50 m^3, built for DM 25 000 showing the highest volumetric gas production (1·56 m^3/m^3 reactor). A more extensive study of 130 digesters (Loll, 1986) indicates that many projects have received help from the Federal Government particularly in construction. Results are reported from monitoring four large projects. These include one plant with two gas-stirred 650 m^3 reactors taking manure from 1100 LU (livestock units, 500 kg weight of fattening bullocks). With a detention period of 10–25 days, the gas represented a net heating value of 43 GJ/day or around one tonne of fuel oil, with the energy being used to provide hot water and electricity (270 kW).

Of 61 farm digesters monitored in France since 1982 (Theoleyre and Heduit, 1987), 24 were continuous systems treating pig slurry and 13 were batch with cattle manure on 50-day cycles. Again gas yields were low (0·2–0·3 m^3 gas/m^3 reactor) as a result of slurry dilution in many systems. However, some continuous processes with thicker slurry showed specific gas yields of over 0·43 m^3 gas/kg TS with a volumetric productivity of 1·3–3·0 m^3 gas/m^3 digester. On average over 85% of the energy content of the gas produced was available for use resulting in some digesters with payback times of less than 6 years, but for 11 farms the payback was estimated as over 10 years.

In Switzerland it has been reported that the cost of the digester reactor may be around 40% of the total for a complete farm system, with gas storage and use 15% and planning 8% of the total costs. A scale of economy was found with an investment cost per animal place of around Sw Fr. 2500 (1981) for 30 head dropping to around Sw Fr. 1500 at 70 head. Annual running costs were about 10% of these values. However, the value of the gas produced fell far short of this, often due to the fact that plants were too large, reflecting the desire of the companies to sell a specific size of tank.

During the 1980s over 30 farm digesters were built in Portugal (Anon., 1989*a*) of which 23 were found to be operating during a survey carried out in 1988, with three more still under construction. Most of the digesters were small (less than 50 m^3), treating manure from herds of pigs of less than 3000 head. This contrasts with Denmark where large scale biogas plants have been promoted (Anon., 1989*b*). The Danish development results from the formation of a

'Coordination Committee for Joint Biogas Plants' following a meeting between the Ministers for Energy, Agriculture and the Environment in Autumn of 1986. The committee was given the task of implementing an action programme which started in early 1988 and will run to the end of 1990. The programme includes studies on existing plant built prior to the initiation of the programme as well as new constructions, a further eight digesters of between 800 and 7200 m^3 so far. The largest, still under construction at Lintrup, will be supplied with around 300 tonnes of manure per day by the 62 farmers who are shareholders in the scheme. This will be supplemented by about 15 tonnes of slaughterhouse waste, 31 tonnes of fish waste and around 10 tonnes of sewage sludge each day. The overall construction cost is expected to be around DKK 44·8 million. Income derived by sale of heat (42 TJ pa) and electricity (4200 MWh/year) from the 11 000 m^3 of gas produced daily is expected to result in an annual cash surplus of about DKK 0·8 million after service of debt. Another similar plant at Ribe is also still under construction. Although slightly smaller in size (5200 m^3) this is expected to have a higher volumetric gas production since it will operate in the thermophilic range. The construction costs estimated at DKK 46 million have been supplemented by government and EC grants of over DKK 17 million.

In Hungary a number of large farm digesters of between 1600 and 2300 m^3 have been built to handle wastes from milk cows or pigs (Banhazi et al., 1986). The largest is a four-reactor system using a complex two-phase aerobic–anaerobic thermophilic/mesophilic process with effluent recycle. The performance of this has been compared with that of a CSTR digester of Austrian design (known as the BIMA digester) and a cost/benefit analysis shows that if over 90% of the gas can be used it is competitive with oil, but as a means of providing heat the systems cannot compete with coal. These and other studies (ADAS/BABA, 1982; Edelmann 1984) stress that where the largest return to the farmer consists of the value of the gas, rather than the benefits from odour and pollution control, then efficiency of gas use is of paramount importance (Table 1.3). The need for low capital cost and therefore a short retention time to produce an economic on-farm digester, as illustrated in Fig. 1.2, has been stressed by Shih (1987). Shih reported the results of a three-year study of thermophilic digestion of chicken manure. High volumetric gas yields (of 4·5 m^3 gas/m^3 digester) were consistently achieved using a semi-plug-flow digester, based on a Taiwanese design. The digester was fed with 6%

Table 1.3 Net present benefits (in £'000) accruing over 10 years discounted at 15% from operation of an anaerobic digester costing £30 000 on a 4000 pig farm. Odour control valued at £2·50 per pig place per annum; based on ADAS/BABA (1982)

Odour control credits Digester performance (% maximum design gas production achieved)	Available				Not available			
	100	60	50	33	100	60	50	33
Used for heating (h/day)								
24	60	42	33	25	9	−7	−16	−25
16	42	30	25	19	−7	−19	−25	−31
12	33	25	20	16	−16	−25	−29	−34
8	25	19	16	13	−25	−30	−34	−36
Used for electricity generation (h/day)								
24	30	17	10	4	−19	−32	−39	−45
16	17	8	4	−0·2	−32	−41	−45	−50
12	10	4	1	−2	−39	−45	−49	−52
8	4	−0	−2	−4	−45	−50	−52	−54
6	1	−2	−3	−5	−49	−52	−54	−89
Gas valued as propane at £0·55/MJ	65	47	38	29	14	−2	−11	−20

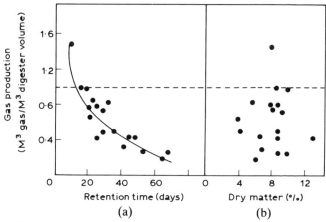

Fig. 1.2 Volumetric gas production in German farm digesters as a function of retention time (a) or feedstock strength (b) expressed as % dry matter. Based on Perwanger (1986). The horizontal line indicates a volumetric gas production of 1 m^3 biogas/m^3 digester volume, the minimum suggested productivity required for an economic system where financial credits associated with pollution or odour control are not assumed.

Table 1.4 Characteristics of a two-stage process for treatment of pig manure (Norrman, 1984)

Fraction	Solids	Liquid
Temperature (°C)	50–52	29–30
Digester	CSTR	Fixed bed
Loading kg VS m^3/day	3·9	4·6
Specific methane yield		
m^3/kg VS	0·15	0·41
m^3/kg COD	0·25	0·36

solids at a five day RT. Other advantages included pollution control, nutrient recycling, pathogen destruction and effluent reuse. These by-products made a substantial contribution to the favourable economics. The capital cost of the system is estimated at around $US1·0 per bird (i.e. $US50 000 for a 50 000 hen operation) with the benefits from the 500 m^3/day biogas depending on the value of the fuel substituted for.

Where pollution control is necessary economics change and more complex systems, with AD as part of an integrated effluent treatment system, are possible. For instance, separated pig wastes (1 tonne TS/day) from a 2200 unit farm in Singapore are treated in a two-stage reactor system aimed at full pollution control (Norrman, 1984). In this plant an anaerobic fixed film reactor is followed by an aerobic trickling filter, clarifier, sand filter and lagoon. Since the slurry is removed by water flushing it is diluted (TS less than 2%) and solids can be removed by settling prior to digestion in a separate stirred tank digester. About 75% of the BOD is removed in the solids digestion with the combined AD part of the system removing 90–96% of the BOD giving an overall gas yield of 0·29 m^3 methane per kg COD removed (Table 1.4).

1.6 Sewage sludge digestion

In the EC mesophilic AD is used to treat about half the sewage sludge produced (80 million wet t equivalent to 3·3 million t dry solids/year) in over 1800 sewage works (Bruce, 1987) generating around 22 million

GJ of energy out of a potential 60 million GJ from this source. Many systems do not reach their potential since the feeds used often contain less than the 3·5% TS required for a positive heat balance. Even so biogas from sewage represents around 30% of the total energy used by the UK water industry.

A survey of sewage digesters in Scotland (Swift and Bruce, 1987) probably reflects the situation in other countries as well. Of 30 heated digesters investigated six had been in service for 50 years, two were recently installed, and the rest were around 20 years old. The most common problems were grit accumulation and corrosion of pipes and gasholders. The most common maintenance activity was replacement of boiler fire tubes. However, in general the digesters functioned satisfactorily with costs of running a 1000 m^3 AD plant £9000 pa of which about half was manpower costs. Surprisingly a number of digesters used considerable amounts of external fuel which would not be necessary if efficiency was improved by removing grit and rags, by thickening feed sludge and by increasing loading rates so more gas was produced.

Older sewage digesters were large concrete structures serving populations of over 100 000 with gas production rates of 3000–6000 m^3/day. Most recent installations, in the UK in particular, have been smaller prefabricated digesters developed from the farm systems of 100–500 m^3 or larger, serving populations of between 1000 and 60 000 with gas production between 50 and 500 m^3/day. The advantages of prefabricated digesters lie in lower construction costs which may be less than 25% of conventional concrete constructions and economic even for small-village-scale plant. For instance in the UK within the Severn–Trent Water Authority the cost of five digesters varying in size from 80 to 1000 m^3 varied from £3 to £6 per person served (Noone, 1987). The interest in prefabricated sewage digesters has been a major factor in the growth of what is now the major AD company in the UK, Farm Gas Ltd (Fig. 1.3).

Although most sewage sludge is treated using stirred tank reactors of one type or another, second generation techniques such as UASB and anaerobic filters are being tested for the treatment of sewage. The operation of a pilot-scale fluidized bed digester processing primary settled domestic sewage has been discussed by Switzenbaum *et al.* (1984) who observed 50% BOD removal with an HRT varying from 1·67 h to 6·67 h.

There is increasing interest in disposing of sewage sludge in

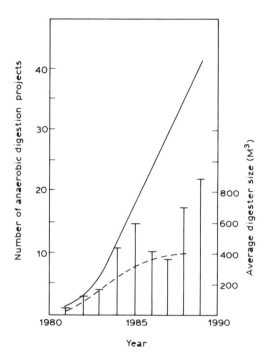

Fig. 1.3 Growth in sales of anaerobic digesters and average digester capacity for projects installed by Farm Gas Ltd, UK (personal communication); —— total number of projects; ----- farm and industrial digesters. Bar chart indicates average size of tanks installed in indicated year.

conjunction with other organic wastes such as MSW by deposition in landfill, in fabricated digesters, or in combination with industrial effluents. The first two alternatives are discussed in the sections on landfill and MSW. An example of the third category has been described by Karlsson (1984) in Sweden where a large (2×3500 m^3) plant run at 40°C handles the combined sewage from 18 000 people and the effluent from a large food processing complex. The plant generates over 1·2 million m^3 of biogas (60% methane) pa and the return in terms of cost benefits and savings in fuel use is around $US0·13 million with an initial investment of $US1·4 million, giving a 10 year payback.

1.7 Industrial effluent digestion

The BOD content of many effluents from the food, fermentation, beverage and paper pulp industries can be reduced by AD treatment. Within the EC Pauss *et al.* (1987) have reported full-scale plant successfully treating material from over 40 different industrial sources. In general terms these effluents may be grouped according to organic load (Brondeau *et al.*, 1984), with low levels (3–10 g/litre) from canning or blanching of vegetables, sugar beet factories, breweries and paper mills, intermediate levels (10–30 g/litre) from pharmaceutical fermentations, distilleries, some breweries, starch factories and cheese production and high concentrations (30–80 g/litre) from chemical process plant.

The various industrial effluents also differ in chemical composition ranging from the sugar factory waters in which the BOD consists mainly of volatile fatty acids in solution, starch effluents from potato and cereal processing which contain colloidal solids to those from canning factories and alcohol production which may contain insoluble particulate material. In all cases the digester has to cope with a high volume of dilute solution making a short HRT essential to reduce capital costs. As a consequence of the high liquid throughput retention of the active biomass in the reactor by some means is also essential. There is therefore widespread use of advanced designs such as contact digesters, UASB plant, anaerobic filters and fluidized bed digesters. Typically these generate between 1000 and 5000 m^3/day of biogas which is usually used in-house as process fuel.

In the UK adoption of industrial AD plant has been limited by a lack of confidence in the technology and a lack of manufacturers of proven systems. This has stimulated a government (SERC) supported demonstration facility in which the performance of four advanced digester types (contact, UASB, filter and fluid bed) is being compared, treating effluent from an ice-cream factory (Anon., 1987*b*). The need for such work in the UK is perhaps surprising since, elsewhere in Europe, the technology has been adopted quite widely. For instance Verrier (1986) cites 22 plants in France, including 14 built since 1983, of which 10 are based on anaerobic filters. Several of the filters, built by SGN (Table 1.5), are downflow reactors packed with a random plastic media giving a high (230 m^2/m^3 reactor) surface area for biomass immobilization. Retention times as low as 8 h give over 90%

Table 1.5 Characteristics of some anaerobic digesters treating industrial effluents (compiled from Demuynck et al., 1984; Biljetina, 1987; Camilleri, 1987 and commercial sources)

Type	Client (Designer)	Influent	Capacity (m^3/day)	Daily load (tonnes COD)	COD reduction (%)	Methane (m^3/m^3 digester)
IRIS	Distillere d'Aiveneux	Distillery waste waters	400	5	80	0.8
Contact	Ets Bondielle	Canning waste water	1 000	10	90–95	0.9
AF	Cellanese Chemical Co.	Chemical process waste	3 784	45	80	1.0
CSTR	NERI	Distillery waste waters	1 200	32	55	1.0
AF	Bacardi Corp	Stillage	1 700	130	65–70	1.5
AF	Revico (SGN)	Stillage	1 920	28	90	3.4
UASB	CSM	Beet sugar factory	1 500	24	75	3.6
AF	Beghin-say (SGN)	Beet waste water	3 000	16	90	4.6

removal of COD at loading rates of up to 20 kg COD m³/day (Camilleri, 1987). The high void volume of the packing used decreases the possibility of clogging by biomass, which is further reduced by expanding the bed slightly by gas recycle to the base of the filter. Information on the performance of a number of French installations is available as a result of monitoring studies carried out by the Agence Française pour la Maîtrise de l'Energie (AFME). These include reports on one of the SGN filters treating residual liquors from a distillery producing Armagnac brandy. The average throughput is 80 m³/h for 8 h/day with a total 150 000 m³ of effluent at 30 g/litre COD (annual load over 4000 t COD with 90% removal) producing 1·4 million m³ of methane per year or 1200 toepa of which 250 toe are used to heat the digester. About half the gas is used to generate electricity and heat the digester as well as greenhouses and the local school in Cognac. The remaining gas is used to provide process heat for wine distillation. The digester replaces a previous stillage concentration system which consumed 3400 toepa, hence the net benefit is equivalent to around 4350 toepa. On a total investment including digester, engineering, assembly, control, civil works, generators, piping, compressors and burners of around FF 20 million the payback time is around three years.

Similar results have been reported for a 1700 m³ anaerobic filter providing an 80% reduction in COD content of potato blanch water at a loading rate of 13 tonnes/day and producing over 4000 m³ of biogas (60–70% methane) (Brondeau et al., 1984). The installation represented a saving of over 400 toepa as a result of closing down aerobic systems and generated biogas worth over 370 toepa which replaced previously bought-in process fuel.

Degremont (France) has publicized details of 25 industrial digesters built worldwide, ranging from small CSTR reactors dealing with 2 t/day COD to a large contact digester handling over 330 t/day COD from distilleries. Results of monitoring studies carried out by AMFE are available for a plant treating 720 m³/day of cannery waste at 28 g/litre COD in two steel digesters of 2500 m³ maintained at 37°C and agitated by gas stirring. The process removes 83% of the applied COD and generates 400 000 m³ pa of gas (53% methane) giving 0·34 m³ of gas per kg COD removed. Annual net gas production is equal to 270 toe whilst 250 toe is saved by not using an aerobic treatment for the same amount of COD. The total financial gain (operating costs less credits for fuel value of the gas and electricity used in an aerobic system) is about FF 0·5 million giving a payback of over eight years at a zero discount rate.

Garcia Buendia (1987) reports 14 pilot- and 11 full-scale industrial systems in operation in Spain. These include UASB reactors of between 1000 and 1600 m^3 for sugar and starch wastes, contact digesters associated with slaughterhouses, paper mills and sugar factories (the largest of 15 000 m^3). Several systems based on anaerobic filters have also been built for the treatment of fish-canning wastes, cheese whey or distillery wastes most of which are around 2000 m^3 in size. Subsidies, from 10–40% of construction costs, are available for installations treating wastes from paper mills, slaughterhouses, sugar industries, distilleries, pharmaceutical plant or breweries.

For the US Biljetina (1987) provides some information listing six large CSTR reactors, variously treating industrial wastes, MSW and cattle manure as well as four anaerobic filters, two fluidized bed reactors, seven plug-flow and three UASB digesters. The list was presented as an example rather than as a summary of all the digesters which have been built in the US. Representative performance data for selected digesters are given in Table 1.5, which includes details of one of the largest fixed film reactors, that built by Bacardi to treat stillage generated during rum manufacture in Puerto Rico (discussed in more detail in Chapter 5).

Another mesophilic (35°C) upflow fixed film installation, based on two reactors each of 100 m^3, built in the US has been described by Love and Roe (1984). This plant treats around 318 m^3/day of potato and other vegetable processing wastewaters with BOD in the range of 1·5–7·5 g/litre resulting in daily loading rates of between 480 and 2380 kg/day. At an HRT of around 16 h 80% of the BOD is removed generating 530 m^3 methane/day of which 340 m^3/day is available for use as process fuel. The digester effluent (2 g/litre BOD) is subsequently treated aerobically to reduce the BOD to 0·2 g/litre which then drops to less than 0·1 g/litre after lagooning. The treated effluent is then used for irrigating a tree plantation. The capital cost was $US224 000 with operating costs of around $US160 per day which are halved if the value of the gas is taken into account. Other benefits of the installation included savings as compared with an equivalent aerobic system (capital $US420 000; operating $US250/day), a 70% decrease in energy requirement, cessation of complaints concerning odour from the factory and a marked decrease in need for lagooning.

Effluent treatment plants based on the UASB principle have been widely adopted by the sugar and starch industries in particular. By 1986 over 60 mesophilic plants of between 30 and 4600 m^3 reactor volume had been built in Europe, the US and Thailand designed to handle between 5 and 20 kg COD m^3/day. More recently over 20 UASB systems have been built in Brazil for treatment of stillage generated by distilleries associated with the National Alcohol Programme (Proalcool), as well as effluents from processing of meat, starch and dairy products (Hirata and Craveiro, 1988). These plants, all of which were commissioned between 1986 and 1989, are large with over two-thirds of the digesters in excess of 1000 m^3. In addition one contact digester system (1800 m^3) and two anaerobic filters have been built.

Although China is best known for its small rural digesters a number of larger industrial digesters have also been built. For instance Zuxuan and Zepeng (1984) describe a two-phase mesophilic (35°C) process, based on anaerobic filter (30 m^3) combined with a UASB (100 m^3) which removed over 70% of the BOD from molasses-derived stillage from an alcohol plant. With a feed COD of 26·5 g/litre and an HRT of less than a day in the first (acidogenic phase) and an HRT of 2·56 days in the second stage the average gas production was around 2·4 volumes of gas per m^3 per day.

1.8 Municipal solid waste digestion

There is increasing interest in using AD as a means of generating biogas from, and reducing the volume of, the organic fraction of MSW. Over the last decade experience has been gained from demonstration plants in the US, Belgium, Italy, France and Japan. A number of commercial plants have been built in France and Italy whilst pilot and laboratory studies have been carried out in the US, UK, Japan, the CEC and elsewhere in Europe (Coombs and Alston, 1988). Experience with these suggests that viable systems will operate at high solids content with gas production highly dependent on the extent and rate of cellulose hydrolysis.

AD of cellulose feedstocks was an important theme of the EC programme on Recycling of Urban and Industrial Waste, which ran from 1979 to the end of 1985 (Ferrero *et al.*, 1984) with several projects investigating the feasibility of anaerobic fermentation of MSW (de Baere & Verstraete, 1984; Pauss *et al.*, 1984; van der Vlugt and

Rulkens, 1984). Pauss et al. (1984) found that digestion of MSW alone in a completely mixed process was unreliable. The pH needed to be adjusted with alkali and the refuse to contain over 40% cellulose or hemicellulose. Other workers however have had greater success. Van der Vlugt and Rulkens (1984) used the fine organic fraction from a mechanical separation plant and obtained a 50% reduction in VS with between 0·4–0·5 m^3 biogas/kg VS at an organic loading rate of 1·85 kg VS/m^3/day, equivalent to 130–160 m^3 of biogas per tonne of fines. The gross costs were estimated at around $US0·46 per m^3 of biogas produced at this loading rate. Extrapolation to a full-scale (100 tonne/day) plant and valuing the gas at $US0·25 per m^3, would mean break-even would occur at an organic loading of around 4 kg VS m^3 digester/day.

De Baere and Verstraete (1984) found that loading rates of 11–13 kg VS/m^3 reactor at retention times of 14–21 days were possible. Between 60–95 m^3 methane/t solids were generated suggesting the process could be economically attractive with a five year payback. The combined value of biogas and compost generated per m^3 of fermenter volume was 200 ECU, for an investment cost of around 1000 ECU/m^3 reactor volume.

Using a small pilot plant (1·2 m^3) fed at a loading rate of 10 kg of sorted MSW m^3/day, with no added water, Klein and Rump (1987) obtained biogas containing 60% methane at a yield of 0·5 m^3/kg TS. Between 50–60% of the gas produced represented a net energy gain. It was anticipated that the throughput could be increased to 60 kg/day with a gas production rate of 5 m^3/m^3 of digester volume at an HRT of about 10 days. A design study has been carried out for a 5 t/day (100 m^3) plant which is expected to produce 500 m^3/day of biogas (55–65% methane) with an energy value of around 11 GJ. On the basis of a pilot scale operation the Arbios company in Belgium (now Organic Waste Systems) are promoting a high solids (30–40%) MSW digestion system (DRANCO) which generates up to 8 m^3 per m^3 reactor volume/day (165 m^3 biogas/tonne DS) plus a compost equal in weight to about 50% of the organic feed (De Baere and Six, 1988). The compost is dried using waste heat from engine cooling if the gas is used for electricity generation.

The largest and longest running full-scale facility is the Refcom plant in Florida, US comprising two 1200 m^3 concrete digesters which have run intermittently since 1978 (Peffer, 1987) but ceased operation in July 1985. Problems related in part to separation and preparation of the feed. The mechanical stirrers caused problems leading to cracking of the concrete roofs of the digesters. Built as an experimental facility it closed once funding was no longer available from the US Department of Energy or the Gas Research Institute which had been supporting the work. However, results from 10 periods of 'steady state' operation are available during which feed rates varied from around 6–18 tonnes per day of mixed sorted/shredded MSW and sewage giving TS of between 2·68 and 6·33% (VS 1·93–4·56%). Higher solids were not possible due to power limitations in the top mounted stirring system, which had blades of 5 m diameter. Gas production varied from 1300–3500 m^3/day with a fairly constant methane content (53%) during various RTs of between 6 and 26 days. There was up to 75% destruction of volatile solids. These results have been used as the basis for a model which suggests that a tipping fee of between $US44 and $US53 would have to be charged for such systems to be viable in the US. This cost is highly dependent on the income which can be generated from sale of methane or electricity (Chynoweth and Legrand, 1988).

The first European demonstration plant treating MSW by anaerobic digestion was built in France by Valorga in 1984. The facility has a capacity of around 16 000 tonnes MSW pa of which 50% passes through the gas-stirred digester. A crushed and screened feed (reduced to 5 cm mesh) diluted to 35–40% solids is fed on a batch basis with an RT of 10–15 days. The digested sludge has a similar composition to conventional composted MSW and is sold as soil conditioner. The net gas production has been reported as 584 000 m^3 pa with around 116 000 m^3 pa being consumed in the plant (Bonhomme and Caire, 1984). Potential gas production rates using this high-solids process are between 4·4 and 5·0 m^3 gas per m^3 reactor volume/day (Begouen et al., 1987). This project has been monitored by AFME who report gas equivalent to 556 toepa sold to the French Gas Board at a rate of 2000 m^3 per day at FF 0·12 per kWh (FF 1390 per toe) with the digested solids sold at FF 80 per tonne. The 1984 investment cost was FF 9·7 million with annual running costs (with a discount rate of 10%, amortization over 15 years and taking into account a 30% investment grant) of over FF 1·0 million even after allowing credits of FF 0·56 million for gas sales and FF 0·26 million for compost sales. The annual deficit can be expressed in terms of a processing cost of around FF 125 per tonne of raw household waste treated.

On the basis of the success of the first demonstration plant at La Buisse a second large installation was built at Amiens, with a capacity of 110 000 tonnes of refuse pa. The construction of this plant, which has three 2400 m³ digesters, was supported by the European Energy Demonstration scheme of the CEC. The company reports that it has orders for the construction of another seven plants with capacities of between 25 000 and 102 000 tpa (Zaoui, 1988).

The digestion of mixtures of sewage sludge and shredded and/or sorted MSW in underground rock caverns has been discussed by Szikriszt (1984) who presents various design and process strategies with RTs varying from years to days. At one extreme the suggested process is equivalent to landfilling whilst at the other extreme rapid digestion of shredded and diluted MSW is similar to that of other fabricated MSW digesters. The rapid digestion concept has been investigated using a 20 m³ pilot reactor with an HRT of 30 days. A mixture of separated waste plus 10–20% sewage sludge diluted to 7% TS at a loading rate of 1·5–1·7 kg VS/m³/day gave methane yields of 0·25–0·29 m³/kg VS.

1.9 Landfill gas

Large amounts of biogas (LFG) are generated spontaneously at rates of between 10 000 and 100 000 m³/day as a result of the disposal of domestic and industrial solid wastes (MSW) by landfilling. Increasingly this gas is being collected in order to reduce problems associated with gas migration and to control offensive odours (ETSU/BABA, 1983; Anon., 1986a; Richards, 1987).

Richards (1987) details about 60 sites in the US where LFG is extracted as well as 10 for which LFG extraction has been considered. The largest, Fresh Kills, New York sells almost 6000 m³ of cleaned gas/h for pipeline use as do about 30% of the sites, most of which have some type of gas clean-up. The earlier sites where LFG production started prior to 1983 usually supply the gas to a local industry for use as a boiler fuel. The most recent sites use the gas to generate electricity in facilities which produce around 0·25–5·5 MW. Capital costs vary from $US0·4 million to $US10 million depending on site size and the extent of gas clean-up. Typically a 75 000 m³/day facility producing medium quality gas would have a return from sale of gas of $US0·8 million giving variable paybacks from less than a year to around three years. Over 40 sites are described in Germany, with about one-third producing over 500 m³ LFG/h with the largest generating over 1500 m³/h. Gas from the sites is used for heating hospitals, residential units, factories or greenhouses. Some sites supply brickworks and 17 sites use the gas to generate electricity which is sold to the public grid.

In the UK there are over 20 landfills producing gas used commercially, or which will start to be used commercially by 1990 (Richards, 1988). The sites which have been exploited for the longest time are those where a local industry was available to use the gas for industrial heating (brick-making, kilns, boilers or horticultural use). More recently several schemes, including those supported in part by the UK Department of Energy, have demonstrated the feasibility of generating electricity using gas engines (1 MW), turbines (3·5 MW) or dual fuel engines (3·0 MW). One of the older, larger (2700 m³/h LFG), monitored sites is at Aveley, gas from which is piped 4 km to a steam boiler in a board mill. The installation cost £267 000 (excluding gas extraction or site work); with gas sold at 15% less than natural gas (28p/therm), an annual saving of about £161 000 which gave a simple payback of around 1·7 years. During 1983/4 the plant achieved over 80% of the design target value supplying gas equivalent to 432 000 GJ or 28% of the total used without any boiler problems being experienced. The average gross calorific value of the gas was around 20 MJ/m³, with a maximum weekly production of 10 000 GJ, dropping following periods of rainfall due to increase in nitrogen content from 8 to 16% (Anon., 1985).

A larger site generating up to 7500 m³ LFG/h at Stone, supplied energy equal to 25 000 tcepa in 1986 which is expected to increase to 35 000 tcepa (Anon., 1986b; Richards, 1987). Construction during 1985/86 cost £2·8 million which, with the anticipated gas production realized, should give a payback of around 3–4 years. The gas is collected by 20 vertical wells 25–30 m deep, connected by a total of 2200 m of pipework, pumped by three compressors rated at 225 KW, with the gas cooled and chilled prior to distribution through a 5 km pipe at a temperature of 2°C to the end-use in three dual fuel cement kiln burners. Several other large sites in the UK do not have suitable industrial users nearby which has led to the interest in supplying the electricity grid. As indicated in Table 1.6 several major projects have

Table 1.6 Electricity generation associated with landfill gas extraction and use in the UK (based on Richards, 1988)

Location	Company	Generation system	Size (MW)	Start year	Energy saving (tcepa)
Aveley	Aveley Methane	Turbine	3·5	1987	7030
Stewartby	Shanks & McEwan	Gas engines	0·8	1987	4420
Meriden	PEEL	Turbine	3·5	1987	14 525
Gerrards Cross	Summerleaze	Dual fuel	3·0	1988	12 450
Mucking	Cory Wastes	—	4·0	1988	16 600
Otterspool	Merseyside Wastes	Gas engine	1·0	1985	4150
Nuneaton	ARC	Gas engine	0·5	1986	2075
Darrington	Darrington Quarries	Duel fuel	1·0	1987	3700
Stone	Blue Circle	Gas engine	1·0	1988	2240
Allsopps	Tarmac Econowastes	Dual fuel	2·0	1988	7855
Preston	LEL/Enercol	Gas engine	2·0	1989	8300

started up over the last 2–3 years and further projects will start up in 1990, giving a total of over 20 MW of electricity being supplied to the grid (Richards, 1988).

There is still an uncertainty concerning how much gas will be produced from a given site and hence gas production is being investigated using simulated landfilling techniques in small test cells. In the US various potential methods of gas enhancement, including accelerated moisture infiltration, elevated moisture content and leachate recycle as well as addition of buffer, sludge and nutrients have been investigated by Walsh *et al.* (1987). After five year's analysis it was concluded that beneficial treatments included sludge, buffer and nutrient addition. Simply increasing the moisture content also had some beneficial effects in the short term by increasing initial gas production rates. In the UK a similar study has been initiated with the construction of six large test cells which will be used to investigate the effects of different placement techniques for the refuse, addition of water and leachate recirculation, addition of sewage sludge, and the mixing together of different wastes. Positive results will further increase the potential of what is, at present, in terms of volumes of biogas produced, the fastest growing sector of the biogas industries on a worldwide basis (Fig. 1.4).

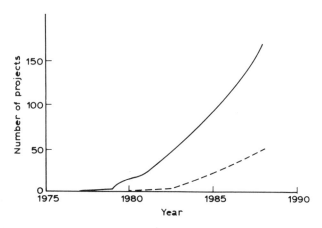

Fig. 1.4 Worldwide growth in number of projects utilizing biogas from landfill sites (based on Richards, 1987). Cumulative number of projects of all types ———; cumulative number of projects generating electricity as the main energy form exported from the site -----.

1.10 Biogas energy systems

In theory large-scale digesters could be built with the sole purpose of supplying pipeline quality gas to a national distribution system using MSW, crop residues, algae or other purpose grown biomass as an extension of experience with both landfill and high solids AD plant. Benson *et al.* (1987) have suggested that a combination of sewage, MSW and crop residues could contribute over 10 EJ of energy by the year 2020 in the US of which 1 EJ could be available by 1995 at a cost of some $US5 billion. The EJ (J raised to the power 18) is roughly equal to the Quad (Btu raised to the power 15), adopted in the US to express large amounts of energy. It is also equal to around 25 mtoe.

The potential gas production from 33 different crops or residues was investigated by Zubr (1984) who obtained the highest yields (0·6 m^3 gas/kg VS) with fresh green material at an RT of less than a month. However, yields dropped to around 0·45 litre/kg VS for straws (wheat or oats), even with much longer RTs of over 80 days emphasizing the problems of lignocellulose digestion. A combination of experiments with a kinetic model led Srivastava and Chynoweth (1987) to conclude that dedicated energy digesters should be designed and operated to optimize solids retention and run at the highest feasible solids concentration. It was noted that effective methods still had to be developed to improve the biodegradability of most forms of biomass, since this was ultimately the limiting factor.

In the US such energy systems could be developed from experimental work carried out at the Community Waste Research Facility at Disney World (Orlando, FL, US) which has the objective of producing large amounts of low-cost pipeline quality methane. Water is purified by being passed through beds of water hyacinth. Biogas is then generated from a mixture of sewage sludge and the sewage-grown biomass digested in a novel high-solids digestion process (Chynoweth *et al.*, 1987*b*; Hayes *et al.*, 1987). The process has been investigated using a 4·5 m^3 vertical flow solids digester handling about 1 tonne wet material per day. At a loading equivalent to 3·2 kg VS m^3/day and with a 2:1 blend of water hyacinth to sludge, steady state conditions gave around 0·4 m^3 gas/kg solids. This was equivalent to a volumetric productivity of around 1·3 vol per vol reactor/day with about 70% solids destruction. The gas yield was increased to around 1·5 vol per vol reactor/day when the feed mode was changed to downflow. If the mix was changed to a 1:1 blend (sludge and water hyacinth) then gas production was 0·45 m^3/kg organic matter added. Both systems were at mesophilic temperatures and at a retention time of 11 days. This corresponded to 92% of the theoretical conversion. Extrapolation of the results suggested that it might be possible to produce pipeline quality gas at a cost of between \$US2 and \$US3 per GJ from alternative lignocellulosic material including plant residues or the organic fraction of MSW.

On the basis of their experience with this system Chynoweth and Isaacson (1987) concluded that large-scale energy digesters are not yet commercial and suggest that AD systems capable of generating biogas in quantities of interest to the gas supply industry will differ significantly from conventional digesters in terms of feed rate, digester unit size, design and method of operation. Important features are simplicity of design and high gas yields on a volumetric basis.

1.11 Conclusions

Technically biogas may be produced from a wide variety of sources. It is not as easy to find the lowest cost, most reliable, optimal system for a given application since many of the factors affecting economic and technical feasibility are site-specific. There are now sufficient companies with a proven track record of design and construction in most sectors, as well as a comprehensive literature detailing past failures, to avoid repetition of past mistakes.

At present energy prices it is unlikely that any AD system will be economic if energy generation is the only reason for construction. However, where AD plants are integrated with other systems in which full value is obtained for all by-products and associated environmental benefits it is now possible to produce an efficient, viable system in most of the sectors discussed above. At present in terms of gas production, no other system can compete with landfilling as an energy source although in future high solids digestion of MSW or large purpose-built energy systems could become important sources of pipeline quality gas. To do this from raw materials high in lignocellulose then pre-treatment, or a two-stage (two-phase) reaction system in which solids breakdown is enhanced may be necessary.

Where the primary objective is to reduce the BOD of industrial effluents a variety of proven technologies is available. The economics depend largely on the local discharge costs to public sewers, the value of gas generated for in-house use and the saving made compared with an alternative (aerobic) disposal system.

Problems with both farm digesters and rural systems often stem from the lack of technical knowledge, quality of design and construction and poor feed. Cost effective systems can be built now, although in developed countries justification has to be linked to environmental protection rather than energy generation.

1.12 References

ADAS/BABA (1982). The Economics of Anaerobic Digestion. Discussion document of an ADAS/BABA working party. Ministry of Agriculture, Fisheries & Food, UK.

Alaa El-Din, M. N. and El-Shumi, S. A. (1986). Small-scale biogas technologies for developing countries: case study Egypt. In *Biomass Conversion for Energy (Biochemical Conversion)*. CNRE bulletin No. 10a. Food & Agricultural Organization of the United Nations, Rome, pp. 69–75.

Anon. (1983). Energy from Biomass in Europe: the 1982 situation. In *Energy from Biomass 2nd EC Conference*, ed. A. Strub, P. Chartier and G. Schleser. Applied Science Publishers, London, pp. 51–62.

Anon. (1984). Number of anaerobic digestion plants reported in Italy as a result of a survey carried out in 1983 by ENEA (Italy). *The Digest*, **11**, 20.

Anon. (1985). The use of landfill gas as a replacement fuel in a water-tube boiler. *Energy Efficiency Demonstration Scheme Final Report* F/60/85/153. Department of Energy, London.
Anon. (1986*a*). The control of landfill gas. In *Waste Management Paper No. 26.* Department of the Environment, HMSO, London, pp. 131–66.
Anon. (1986*b*). Blue Circle Industries plc The Stone Project. *The Digest,* **16**, 20.
Anon. (1987*a*). *Review,* Issue 1. Department of Energy, London.
Anon. (1987*b*). SERC anaerobic digestion research facility. *The Digest,* **18**, 6.
Anon. (1989*a*). *Biogas em Portugal.* Direccao-Geral de Energia, Lisbon.
Anon. (1989*b*). *Large-scale Biogas Plants.* Danish Energy Agency, Kobenhavn.
Banhazi, G., Flieg, J. and Velez, D. (1986). Operational experience with large scale biogas plants in Hungary. In *Biomass Conversion for Energy (Biochemical Conversion). Proceedings of a Technical Consultation (October 1985).* CNRE bulletin No. 10a. Food and Agricultural Organization of the United Nations, Rome, pp. 63–8.
Begouen, O., Pavia, A., Thiebault, E. and Peillex, P. J. (1987). Continuous high solids content methanization of a polysubstrate mixture of municipal solid waste and sludge. In *Biomass for Energy and Industry 4th EC Conference,* ed. G. Grassi, B. Delmon, J.-F. Molle and H. Zibetta. Elsevier Applied Science Publishers, London, pp. 927–34.
Benson, P. H., Hayes, T. D. and Isaacson, R. (1987). Regional and community approaches to methane from biomass and waste: an industry perspective. In *Energy from Biomass and Wastes X,* ed. D. L. Klass. Elsevier Applied Science Publishers, London and Institute of Gas Technology, Chicago, pp. 987–1008.
Biljetina, R. (1987). Commercialization and economics. In *Anaerobic Digestion of Biomass,* ed. D. P. Chynoweth and R. Isaacson. Elsevier Applied Science Publishers, London, pp. 231–55.
Bonazzi, G., Cortellini, L., Piccinini, S. and Tilche, A. (1984). The bio-gas project in Emilia-Romagna (Italy). In *Bioenergy 84,* ed. H. Egneus and A. Ellegard. Elsevier Applied Science Publishers, London, pp. 333–8.
Bonhomme, M. and Caire, B. (1984). Recycling of urban waste by a continuous highly concentrated fermentation process producing methane gas and soil conditioner. In *Bioenergy 84, Vol III Biomass Conversion,* ed. H. Egneus and A. Ellegard. Elsevier Applied Science Publishers, London, pp. 366–9.
Brondeau, P., Mouliney, M. and Cutayar, J. (1984). The Anoxal process: treatment of liquid industrial effluents by means of anaerobic filters. In *Bioenergy 84, Vol III Biomass Conversion,* ed. H. Egneus and A. Ellegard. Elsevier Applied Science Publishers, London, pp. 387–91.
Bruce, A. M. (1987). Progress on anaerobic digestion within the framework of the EC concerted action cost 681. In *Anaerobic Digestion: Results of Research and Demonstration Projects,* ed. M. P. Ferranti, G. L. Ferrero and P. L'Hermite. Elsevier Applied Science Publishers, London, pp. 177–96.

Caceres, R. and Chiliquinga, B. (1986). Experiences with rural digesters in Latin America. In *Biogas Technology: Transfer and Diffusion*, ed. M. M. El-Halwagi. Elsevier Applied Science Publishers, London, pp. 150–65.
Camilleri, C. (1987). Operating results from a fixed-film anaerobic digester for pollution abatement and methane production from industrial wastes. In *Biomass for Energy and Industry 4th EC Conference*, ed. G. Grassi, B. Delmon, J.-F. Molle and H. Zibetta. Elsevier Applied Science Publishers, London, pp. 1338–42.
Chantavorapap, S. (1986). Biogas program of Thailand. In *Biogas Technology: Transfer and Diffusion*, ed. M. M. El-Halwagi. Elsevier Applied Science Publishers, London, pp. 691–4.
Chesshire, M. J. (1986). A comparison of the design and operational requirements for the anaerobic digestion of animal slurries and of sewage sludge. In *Anaerobic Digestion of Sewage Sludge and Organic Agricultural Wastes*, ed. A. M. Bruce, A. Kouzeli-Katsiri and P. J. Newman. Elsevier Applied Science Publishers, London, pp. 33–54.
Chynoweth, D. P. and Isaacson, R. (eds) (1987). *Anaerobic Digestion of Biomass*. Elsevier Applied Science Publishers, London.
Chynoweth, D. P. and Legrand, R. (1988). Anaerobic digestion as an integral part of municipal solid waste management. In *Landfill Gas and Anaerobic Digestion of Solid Waste*, ed. Y. R. Alston and G. E. Richards. UK Department of Energy, London, pp. 467–80.
Chynoweth, D. P., Fannin, K. F. and Srivastava, V. J. (1987a). Biological gasification of marine algae. In *Seaweed Cultivation for Renewable Resources: Developments in Aquaculture and Fisheries Science Volume 16*, ed. K. T. Bird and P. H. Benson. Elsevier Science Publishers BV, Amsterdam, pp. 285–303.
Chynoweth, D. P., Biljetina, R., Srivastava, V. and Hayes, T. (1987b). A novel solids-concentrating anaerobic digester. In *Biomass for Energy and Industry 4th EC Conference*, ed. G. Grassi, B. Delmon, J.-F. Molle and H. Zibetta. Elsevier Applied Science Publishers, London, pp. 828–33.
Constant, M., Naveau, H., Ferrero, G.-L. and Nyns, E.-J. (1989). *Biogas, End-Use in the European Community*. Elsevier Applied Science, London.
Coombs, J. and Alston, Y. R. (1988). The potential for recovery of energy from solid wastes using fabricated digester systems. In *Landfill Gas and Anaerobic Digestion of Solid Waste*, ed. Y. R. Alston and G. E. Richards. UK Department of Energy, London, pp. 454–66.
de Baere, L. and Six, W. (1988). Dry anaerobic digestion of agro-industrial wastes. In *Landfill Gas and Anaerobic Digestion of Solid Waste*, ed. Y. R. Alston and G. E. Richards. UK Department of Energy, London, pp. 545–9.
de Baere, L. and Verstraete, W. (1984). Anaerobic fermentation of semi-solid and solid substrates. In *Anaerobic Digestion and Carbohydrate Hydrolysis of Waste*, ed. G. L. Ferrero, M. P. Ferranti and H. Naveau. Elsevier Applied Science Publishers, London, pp. 195–208.
Deming, J. W. (1987). Potential for bacterial methane production at superheated temperatures. In *Energy from Biomass and Wastes X*, ed. D. L.

Klass. Elsevier Applied Science Publishers, London and Institute of Gas Technology, Chicago, pp. 1097–113.
Demuynck, M., Nyns, E.-J. and Palz, W. (1984). *Biogas Plants in Europe Solar Energy R & D in the European Community Series E Volume 6*. D. Reidel Publishing Co., Dordrecht.
Durand, J. H., Fischer, J. R., Jannotti, E. L. and Miles, J. B. (1987). Development of an economic system model to optimize the performance of a swine waste digestion system. *Biomass*, **14** 219–44.
ETSU/BABA (1983). *Landfill Gas Workshop. Proceedings*. Harwell Laboratory and the British Anaerobic and Biomass Association, UK.
Edelmann, W. E. (1984). The feasibility of biogas production on modern farms. In *Bioenergy 84 Vol I Bioenergy State of the Art*, ed. H. Egneus and A. Ellegard. Elsevier Applied Science Publishers, London, pp. 297–304.
Edelmann, W. E. (1986). Technologies for biogas production of developed countries and possibilities of transferring them to developing countries. *Biogas Technology: Transfer and Diffusion*, ed. M. M. El-Halwagi. Elsevier Applied Science Publishers, London, pp. 204–12.
El-Halwagi, M. M. (ed.) (1986). *Biogas Technology: Transfer and Diffusion*. Elsevier Applied Science Publishers, London.
Ferrero, G. L., Ferranti, M. P. and Naveau, H. (eds) (1984). *Anaerobic Digestion and Carbohydrate Hydrolysis of Waste*. Elsevier Applied Science Publishers, London.
Friman, R. M. (1986). Anaerobic digestion on farms in the United Kingdom. In *Anaerobic Digestion of Sewage Sludge and Organic Agricultural Wastes*, ed. A. M. Bruce, A. Kouzeli-Katsiri and P. J. Newman. Elsevier Applied Science Publishers, London, pp. 135–44.
Garcia Buendia, A. J. (1987). Biogas technology developed and evaluated in Spain. In *Biomass for Energy and Industry 4th EC Conference*, ed. G. Grassi, B. Delmon, J.-F. Molle and H. Zibetta. Elsevier Applied Science Publishers, London, pp. 289–94.
Gorkhali, H. G. (1986). Summary of the Nepal biogas program. In *Biogas Technology: Transfer and Diffusion*, ed. M. M. El-Halwagi. Elsevier Applied Science Publishers, London, pp. 665–8.
Hall, E. R. and Hobson, P. N. (eds) (1988). *Anaerobic Digestion 1988*. Pergamon Press, Oxford.
Havelaar, A. H. (1985). Conclusions. In *Inactivation of Microorganisms in Sewage Sludge by Stabilisation Processes*, ed. D. Strauch. A. H. Havelaar and P. L'Hermite. Elsevier Applied Science Publishers, London, pp. 189–90.
Hayes, T. D., Isaacson, H. R., Chynoweth, D. P., Reddy, K. R. and Biljetina, R. (1987). An integrated wastewater-energy production system. In *Energy from Biomass and Wastes X*, ed. D. L. Klass. Elsevier Applied Science Publishers, London and Institute of Gas Technology, Chicago, pp. 1127–48.
Hirata, Y. S. and Craveiro, A. M. (1988). Estagio de desenvolvimento e aplicacao da digestao anaerobia no Brasil. In *Proceedings of Seminar on Energy and Wastes*, April 1988, DG XVII, CEC, Brussels.

Karlsson, P. O. (1984). Esloev biogas plant two years of operation. In *Bioenergy 84, Vol III Biomass Conversion*, ed. H. Egneus and A. Ellegard. Elsevier Applied Science Publishers, London, pp. 354–6.

Klein, M. and Rump, H. (1987). Anaerobic digestion of solids: organic fraction of municipal solid waste (MSW). In *Biomass for Energy and Industry 4th EC Conference*, ed. G. Grassi, B. Delmon, J.-F. Molle and H. Zibetta. Elsevier Applied Science Publishers, London, pp. 845–54.

Loll, U. (1986). Biogas plants for animal slurries in the Federal Republic of Germany. In *Anaerobic Digestion of Sewage Sludge and Organic Agricultural Wastes*, ed. A. M. Bruce, A. Kouzeli-Katsiri and P. J. Newman. Elsevier Applied Science, London, pp. 14–32.

Love, S. and Roe, S. F., Jr (1984). Fixed film anaerobic digestion on a commercial scale for potato and vegetable wastes. In *Bioenergy 84, Vol III Biomass Conversion*, ed. H. Egneus and A. Ellegard. Elsevier Applied Science Publishers, London, pp. 378–86.

Moulik, T. K., Singh, J. B. and Vyas, S. K. (1986). The biogas program in India. In *Biogas Technology: Transfer and Diffusion*, ed. M. M. El-Halwagi. Elsevier Applied Science Publishers, London, pp. 661–4.

Mowat, D. N., Jones, C. R., Buchanan-Smith, J. G. and Macleod, G. K. (1986). Nutritive value of methane fermentation residue produced from cattle and swine wastes. In *Microbial Biomass Proteins*, ed. M. Moo-Young and K. F. Gregory. Elsevier Applied Science Publishers, London, pp. 167–73.

Nielsen, V. C., Voorburg, J. H. and L'Hermite, P. (eds) (1986). *Odour Prevention and Control of Organic Sludge and Livestock Farming*. Elsevier Applied Science Publishers, London.

Noone, G. P. (1987). Anaerobic sludge treatment of urban wastes. In *Proceedings of a WRc/BABA workshop on Prefabricated Anaerobic Sludge Digesters*. WRc Processes, Stevenage and British Anaerobic and Biomass Association, Reading, UK, pp. 24–53.

Norrman, J. (1984). Anaerobic treatment of piggery waste—thermophilic digestion of the solid phase and mesophilic treatment of the liquid phase. In *Bioenergy 84, Vol III Biomass Conversion*, ed. H. Egneus and A. Ellegard. Elsevier Applied Science Publishers, London, pp. 301–9.

Orcullo, N. A. (1986). Biogas technology development and diffusion: the Philippine experience. In *Biogas Technology: Transfer and Diffusion*, ed. M. M. El-Halwagi. Elsevier Applied Science Publishers, London, pp. 669–85.

Pain, B. F., West, R., Oliver, B. and Hawkes, D. L. (1984). Mesophilic anaerobic digestion of dairy cow slurry on a farm scale: first comparisons between digestion before and after solids separation. *Journal of Agricultural Engineering Research*, **29**, 249–56.

Pauss, A., Nyns, E.-J. and Naveau, H. (1984). Production of methane by anaerobic digestion of domestic refuse. In *Anaerobic Digestion and Carbohydrate Hydrolysis of Waste*, ed. G. L. Ferrero, M. P. Ferranti and H. Naveau. Elsevier Applied Science Publishers, London, pp. 209–22.

Pauss, A., Mahy, D., Schepens, G., Naveau, H., Palz, W. and Nyns, E.-J. (1987). Biogas plants in Europe: an updated data bank. In *Biomass for*

Energy and Industry 4th EC Conference, ed. G. Grassi, B. Delmon, J.-F. Molle and H. Zibetta. Elsevier Applied Science Publishers, London, pp. 914–19.
Peffer, J. T. (1987). Evaluation of the Refcom proof-of-concept experiment. In *Energy from Biomass and Wastes X*, ed. D. L. Klass. Elsevier Science Publishers, London and Institute of Gas Technology, Chicago, pp 1149–71.
Pernkopf, J. (1986). Biogas technology in Austria—Potential economic aspects, investigations for improvement. In *Biomass Conversion for Energy (Biochemical Conversion)*, CNRE bulletin No. 10a. Food and Agricultural Organization of the United Nations, Rome, pp. 61–2.
Perwanger, A. (1986). Experiments with demonstration biogas plants. In *Biomass Conversion for Energy (Biochemical Conversion)*. CNRE bulletin No. 10a, Food and Agricultural Organization of the United Nations, Rome, pp. 61–2.
Pluschke, P. (1986). Analysis of economic factors in the dissemination of biogas plants—case studies from Africa and the Caribbean. In *Biogas Technology: Transfer and Diffusion*, ed. M. M. El-Halwagi. Elsevier Applied Science Publishers, London, pp. 120–31.
Richards, K. M. (1987). Landfill gas—a global review. In *Proceedings Seventh International Biodeterioration Symposium*, ETSU L-22. AERE, Harwell, UK.
Richards, K. M. (1988). The UK landfill gas and MSW industry so far, so good? In *Landfill Gas and Anaerobic Digestion of Solid Waste*, ed. Y. R. Alston and G. E. Richards. UK Department of Energy, London, pp. 12–46.
Shih, J. C. H. (1987). Ecological benefits of anaerobic digestion. *Poultry Science*, **66**, 946–50.
Smil, V. (1986). The realistic potential of biogas. In *Biogas Technology: Transfer and Diffusion*, ed. M. M. El-Halwagi. Elsevier Applied Science Publishers, London, 45–50.
Srivastava, V. J. and Chynoweth, D. P. (1987). Kinetic analysis of biogasification of biomass/waste blend and its engineering significance. In *Energy from Biomass and Wastes X*, ed. D. L. Klass. Elsevier Applied Science Publishers, London and Institute of Gas Technology, Chicago, pp. 1021–34.
Stuckey, D. C. (1986). Biogas: A global perspective. In *Biogas Technology: Transfer and Diffusion*, ed. M. M. El-Halwagi. Elsevier Applied Science Publishers, London, pp. 18–44.
Swift, D. W. and Bruce, A. M. (1987). *A Survey of Heated Anaerobic Digester Operation at Sewage Works in Scotland*. WRc Processes, Stevenage.
Switzenbaum, M. S., Sheehan, K. C. and Hickey, R. F. (1984). Anaerobic processes for municipal wastewater. In *Bioenergy 84, Vol III Biomass Conversion*, ed. H. Egneus and A. Ellegard. Elsevier Applied Science Publishers, London, pp. 348–53.
Szikriszt, G. (1984). The WBG method. In *Bioenergy 84, Vol III Biomass Conversion*, ed. H. Egneus and A. Ellegard. Elsevier Applied Science Publishers, London, pp. 370–7.

Theoleyre, M. A. and Heduit, M. (1987). Biomethanation of agricultural wastes in France. In *Biomass for Energy and Industry 4th EC Conference*, ed. G. Grassi, B. Delmon, J.-F. Molle and H. Zibetta. Elsevier Applied Science Publishers, London, pp. 289–94.

Van der Vlugt, A. J. and Rulkens, W. H. (1984). Biogas production from a domestic waste fraction. In *Anaerobic Digestion and Carbohydrate Hydrolysis of Waste*, ed. G. L. Ferrero, M. P. Ferranti and H. Naveau. Elsevier Applied Science Publishers, London, pp. 245–50.

Verrier, D. (1986). Méthanisation d'effluents industriels en fermenteurs à film fixe. *Entropie*, **130/131**, 35–8.

Walsh, J. J., Stamm, J. W., Vogt, W. G., Kinman, R. N. and Rickabaugh, J. I. (1987). Demonstration of landfill gas enhancement techniques in landfill simulators. In *Energy from Biomass and Wastes X*, ed. D. L. Klass. Elsevier Applied Science Publishers, London and Institute of Gas Technology, Chicago, pp. 1115–25.

Ward, R. F. (1986). Engineering design of biogas units for developing countries. In *Biogas Technology: Transfer and Diffusion*, ed. M. M. El-Halwagi. Elsevier Applied Science Publishers, London, pp. 178–203.

Zaoui, R. (1988). The Valorga digestion process. In *Landfill Gas and Anaerobic Digestion of Solid Waste*, ed. Y. R. Alston and G. E. Richards. UK Department of Energy, London, pp. 481–500.

Zubr, J. (1984). Biogas energy potentials of energy crops and crop residues. In *Bioenergy 84, Vol III Biomass Conversion*, ed. H. Egneus and A. Ellegard. Elsevier Applied Science Publishers, London, pp. 295–300.

Zuxuan, W. and Zepeng, C. (1984). Disposing alcohol from molasses sewage using a new type anaerobic digester. In *Bioenergy 84, Vol III Biomass Conversion*, ed. H. Egneus and A. Ellegard. Elsevier Applied Science Publishers, London, pp. 361–5.

2 The microbiology and control of anaerobic digestion

D. B. Archer and B. H. Kirsop

AFRC Institute of Food Research, Norwich, UK

2.1	**Introduction**	44
2.2	**Microbiology of anaerobic digesters**	44
2.2.1	Functional groups of bacteria	45
2.2.2	Hydrolysis and acidogenesis	45
2.2.3	The acetogenic bacteria	46
2.2.4	The methanogens	48
2.3	**The controlling reactions in anaerobic digestion**	48
2.3.1	Insoluble substrates	50
2.3.2	Soluble substrates	51
2.3.3	Methanogenesis from acetate	52
2.3.4	The role of hydrogen	56
2.4	**Metabolic interactions of methanogens with other organisms**	56
2.4.1	Neutralism	57
2.4.2	Commensalism	57
2.4.3	Mutualism	58
2.4.4	Predator–prey and amensalism	60
2.4.5	Competition	60
2.5	**Nutrition of the methanogenic bacteria**	62
2.5.1	Trace organic nutrients	62
2.5.2	Nitrogen requirements	63
2.5.3	Phosphate requirements	64
2.5.4	Sulphur requirements	65
2.5.5	The effects of metals	65
2.6	**Waste treatment by anaerobic mixed cultures**	67
2.6.1	Monitoring anaerobic digestion	70
2.6.2	Control of nutrients	71
2.6.3	Monitoring hydrogen	72
2.7	**Conclusions**	73
2.8	**References**	74

2.1 Introduction

The anaerobic microbial degradation of organic matter to methane and carbon dioxide occurs naturally in a variety of anaerobic habitats such as intestinal tracts and sediments. The microorganisms are exploited in the biotechnological process of anaerobic digestion both to reduce the pollution caused by organic wastes and to produce methane which can be used as a fuel. Anaerobic digestion of a waste prior to release into a natural water course, for example, offers a means of reducing the polluting effects of the waste. The economic disposal of waste and legislation designed to protect the environment from pollution necessitates simple and effective methods of waste treatment. Anaerobic digestion has the potential to be an economically attractive process. Examples given elsewhere in this book show that, in some cases, it has already fulfilled its potential. Equally, however, the failure of many anaerobic digesters to operate reliably at their design performance has underlined the need for more basic information on the biological aspects of anaerobic digestion (Dunnill and Rudd, 1984). There is a need not only for more information on the biology of the system but information is required on how to exploit this to control the anaerobic digestion process. Control of the process is necessary in landfills, animal slurry digesters and industrial effluent digesters. In landfills reliable and constant gas production is needed over a period of many years if incorporating a gas abstraction system into the landfill is to be economic. At the end of this period of gas production, ideally, the landfill should be inactive and the gas exhausted so that the land can be usefully reclaimed. The economics of gas production from agricultural wastes are less favourable than from industrial effluents for there is often no charge for pollution and the economics then depend, crucially, upon the value of the methane. With the treatment of industrial effluents the reduction in effluent charges arising from incomplete treatment is usually more valuable than the methane produced. The reliable conversion of organic matter to methane, preferably at high rate, is needed and this requires careful control of the mixed cultures which convert the wastes to methane.

2.2 Microbiology of anaerobic digesters

The numbers and types of microorganisms present in digesters are likely to depend upon the type of digester, its operating conditions and the waste composition. Few studies have been made of the microbiology of landfills, and even then, studies have mainly been restricted to microbial activities or to the more easily cultivated bacteria (Cromwell, 1965; Filip and Küster, 1979; Jones and Grainger, 1983; Jones et al., 1983; Senior and Balba, 1983). The microbiology of landfills is probably similar, although not necessarily in detail, to that of other anaerobic digesters because in all cases organic material is converted to the same products. There are, however, differences in the microbiology of the rumen and agricultural waste digesters (Hobson et al., 1974; Hobson, 1981) and the operating conditions of digesters affect methanogenic populations (Smith et al., 1980). There is no comprehensive analysis of the microbial composition in terms of species present, numbers and their organisation, in any anaerobic digester. Knowledge of the differences in microbial composition between digester types and digesters operated under varying conditions is limited and it is therefore only possible to draw some general conclusions as to the species most likely to be present. As will be seen the bacterial species can be categorised into functional groups and investigations on the detailed microbiology can be restricted, at least at the present state of knowledge, to those groups which exert a controlling effect on the degradation of wastes.

2.2.1 Functional groups of bacteria

The metabolic stages involved in the production of methane from wastes (Fig. 2.1) are hydrolysis, acidogenesis, acetogenesis and methanogenesis (Bryant, 1979; Kirsop, 1984). Schemes for the metabolic conversions in methanogenesis are extremely complicated but even so are not yet a complete description of the process. Often the complexity only serves to hide the essential features. Nevertheless, in terms of the controlling reactions in anaerobic digestion the scheme in Fig. 2.1 can be simplified further to become a balance between the production and removal of electrons (e.g. as H_2 or formate) and H^+. This aspect is discussed in detail below.

2.2.2 Hydrolysis and acidogenesis

The hydrolytic and acidogenic stages may be combined in the anaerobic acidogenic bacteria. Acidogenic bacteria which are com-

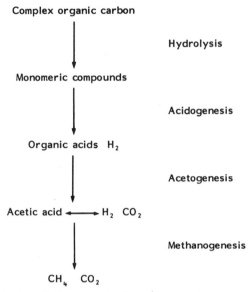

Fig. 2.1 Metabolic stages in the anaerobic digestion of wastes.

monly found in digesters include species of *Butyrivibrio*, *Propionibacterium*, *Clostridium*, *Bacteroides*, *Ruminococcus*, *Acetivibrio*, *Bifidobacterium*, *Eubacterium*, *Peptostreptococcus*, *Peptococcus*, *Selenomonas*, *Lactobacillus*, *Streptococcus* and members of the *Enterobacteriaceae* (Zeikus, 1980; Hobson, 1981; Iannotti *et al.*, 1982; Sahm, 1984) although this is by no means an exhaustive list. In mesophilic sewage sludges there are usually between 10^8–10^9 hydrolytic bacteria per ml (Toerien *et al.*, 1967).

2.2.3 The acetogenic bacteria

Acetogenic species can be subdivided into those which are not obligately proton-reducing, i.e. hydrogen-producing, species and those which do reduce protons to hydrogen obligately during acetogenesis. The first group is broad, including the homoacetogens and species which may direct their metabolism to proton-reduction in the presence of an efficient hydrogen-removing system. Homoacetogenic species are known in the genera *Acetobacterium* (Balch *et al.*, 1977; Braun and

Gottschalk, 1982; Eichler and Schink, 1984), *Acetoanaerobium* (Sleat *et al.*, 1985), *Acetogenium* (Leigh and Wolfe, 1983), *Butyribacterium* (Zeikus *et al.*, 1980), *Clostridium* (Wieringa, 1940; Ohwaki and Hungate, 1977; Adamse, 1980; Wiegel *et al.*, 1981; Schink, 1984*a*), *Eubacterium* (Sharak-Genthner *et al.*, 1981) and *Pelobacter* (Schink and Pfennig, 1982; Schink, 1984*b*). In mesophilic sludges there are approximately 10^5 homoacetogens per ml forming acetate from $H_2 + CO_2$ (Braun *et al.*, 1979). The competitive ability of these homoacetogens for H_2 in mixed cultures has not yet been clearly established but they can form stable mutualistic associations (see Table 2.2) with H_2-producing bacteria and one thermophilic mutualistic coculture has been isolated. It grew under slightly acidic conditions (Le Ruyet *et al.*, 1984*a*). In environments with efficient H_2 sinks, such as anaerobic digesters, many of the acidogenic species direct their metabolism to acetogenesis. This facultative change in metabolism has been demonstrated in defined methanogenic cocultures degrading alcohols, lactate, pyruvate, cellobiose, glucose, fructose and cellulose (Mah, 1982).

Obligately proton-reducing acetogenic bacteria can only be grown in an efficient electron-removing environment, for example in monoxenic culture with a hydrogen-removing or formate-removing species. The simplest mixed culture involving this type of 'mutualistic' interaction is a culture containing the acetogen and a hydrogen-removing bacterium such as a methanogen. The first described example of mutualism, where hydrogen is the metabolic link between the two species, was the coculture of the so called S organism with *Methanobacterium bryantii* (Bryant *et al.*, 1967). This had previously been described as the single species *Methanobacterium omelianskii* (Barker, 1940) renamed later as *Methanobacillus omelianskii* (Barker, 1956). The S organism could grow axenically when metabolising pyruvate (Reddy *et al.*, 1972) but depended upon the methanogen to facilitate its proton-reductive metabolism of ethanol. In the absence of sulphate *Desulfovibrio* spp. are obligate proton-reducing acetogens when metabolising ethanol or lactate and can be grown in mutualistic coculture with methanogens (Bryant *et al.*, 1977; McInerney and Bryant, 1981). Other obligate proton-reducing acetogens have been described: *Syntrophobacter wolinii* degrades propionate (Boone and Bryant, 1980), *Syntrophomonas wolfei* degrades butyrate (McInerney *et al.*, 1979, 1981), *Syntrophus buswellii* degrades benzoate (Mountfort and Bryant, 1982; Mountfort *et al.*, 1984). Conditions may yet be found which support the growth of the acetogens in axenic culture. For example, *Syntrophococcus sucro-*

mutans grows well in mutualistic coculture with a methanogen but some organic electron acceptors support its growth in pure culture (Krumholz and Bryant, 1986). A thermophilic butyrate-degrading species growing in association with *Methanobacterium thermoautotrophicum* has been described (Henson and Smith, 1985). Methanogenesis from long chain fatty acids (Heukelekian and Mueller, 1958; Chynoweth and Mah, 1971; Roy *et al.*, 1985) and 3-chlorobenzoic acid (Shelton and Tiedje, 1984) is also thought to involve obligately proton-reducing acetogenic bacteria. McInerney *et al.* (1979) reported that in mesophilic digester sludge *S. wolfei* was present at a level of at least $4 \cdot 5 \times 10^6$ per g. The concept of inter-species formate transfer has also recently been described (Thiele and Zeikus, 1988; Boone *et al.*, 1989). The relative importance of hydrogen and formate in inter-species electron transfer needs to be established in different digesters and under different operating conditions.

2.2.4 The methanogens

Methanogens are present in sewage sludges at populations up to 10^8 per ml (Smith, 1966) and contribute up to 10% of the volatile solids (van Beelen *et al.*, 1983a). They are a morphologically diverse group of archaebacteria unified by their ability to derive energy from methanogenesis (Balch *et al.*, 1979). A limited range of substrates are utilised by the methanogens, acetate and $H_2 + CO_2$ being the most important substrates in anaerobic digestion. Table 2.1 lists the named methanogenic bacteria together with substrates utilised and indicates which have been found in anaerobic digesters. Most methanogenic bacteria utilise $H_2 + CO_2$ but species of only two genera, *Methanosarcina* and *Methanothrix*, can produce methane from acetic acid. A more comprehensive list of methanogens is now available (Boone and Whitman, 1988).

2.3 The controlling reactions in anaerobic digestion

The production of methane from wastes involves a complex mixed culture. One object of research into the microbiology of anaerobic digestion is to improve control of the process and it is necessary to

The microbiology and control of anaerobic digestion 49

Table 2.1 Methanogenic species described up to and including 1986*

Species	Methanogenic substrates	Found in digesters, +/−
Methanobacterium		
bryantii	$H_2 + CO_2$	+
formicicum	$H_2 + CO_2$, formate	+
wolfei	$H_2 + CO_2$	+/−
thermoautotrophicum	$H_2 + CO_2$	+
uliginosum	$H_2 + CO_2$	−
thermoalcaliphilum	$H_2 + CO_2$	+
thermoaggregans	$H_2 + CO_2$	−
Methanobrevibacter		
arboriphilus	$H_2 + CO_2$	+
ruminantium	$H_2 + CO_2$, formate	+
smithii	$H_2 + CO_2$, formate	+
Methanothermus		
fervidus	$H_2 + CO_2$	−
Methanococcus		
maripaludis	$H_2 + CO_2$, formate	−
deltae	$H_2 + CO_2$	−
vannielii	$H_2 + CO_2$, formate	−
voltae	$H_2 + CO_2$, formate	−
jannaschii	$H_2 + CO_2$, formate	−
halophilus	Methanol, methylamines	−
thermolithotrophicus	$H_2 + CO_2$, formate	−
frisius	$H_2 + CO_2$, methanol, methylamines	−
Methanomicrobium		
mobile	$H_2 + CO_2$, formate	−
paynteri	$H_2 + CO_2$	−
Methanogenium		
cariaci	$H_2 + CO_2$, formate	−
marisnigri	$H_2 + CO_2$, formate	−
olentangyi	$H_2 + CO_2$	−
tatii	$H_2 + CO_2$, formate	−
aggregans	$H_2 + CO_2$, formate	+
thermophilicum	$H_2 + CO_2$, formate	−
bourgense	$H_2 + CO_2$, formate	+
Methanospirillum		
hungatei	$H_2 + CO_2$, formate	+
Methanoplanus		
limicola	$H_2 + CO_2$, formate	−

(continued)

Table 2.1—*contd.*

Species	Methanogenic substrates	Found in digesters, +/−
Methanosarcina		
barkeri	⎧ $H_2 + CO_2$, acetate	+
mazei	⎨ methanol, methlamines	+
acetivorans	Acetate, methanol, methylamines	−
TM1	Acetate, methanol, methylamines	+
Methanococcoides		
methylutens	Methanol, methylamines	−
Methanothrix		
soehngenii	Acetate	+
concilii	Acetate	+
Methanolobus		
tindarius	Methanol, methylamines	−
Methanosphaera		
stadtmaniae	H_2 + methanol	−

* Refer to Boone and Whitman (1988) for a more up to date list of methanogenic species. Issues of the *International Journal of Systematic Bacteriology* since April 1988, should also be consulted.

decide which parts of the interdependent complex processes are limiting and are amenable to control. The concept of the rate-limiting step focuses attention upon a single part of the process but of the four metabolic stages illustrated in Fig. 2.1 only the acidogenic stage has not at some time been claimed to be rate-limiting (Archer, 1984*a*; Archer *et al.*, 1987).

2.3.1 *Insoluble substrates*

Hydrolysis is claimed to be rate-limiting when the waste contains much insoluble material (Eastman and Ferguson, 1981; Boone, 1982). Volatile acids may still accumulate during the degradation of insoluble matter, and this is common during anaerobic digestion in landfills (Rees, 1980; Harmsen, 1983). The major cellulolytic bacteria in anaerobic digesters are species of *Clostridium*, *Ruminococcus*, *Butyrivibrio* and *Bacteroides* (Hobson *et al.*, 1974; Le Ruyet *et al.*, 1984*b*; Veal and Lynch, 1984). Experimentation with model cellulose substrates, although avoiding practical difficulties of degrading the cellulose in ligno-cellulose, has revealed many details of mixed-culture interactions in cellulolysis (Mah, 1982; Veal and Lynch, 1984). Methanogens affect cellulolysis by removing H_2 and acetic acid (Khan, 1980; Mah, 1982) and other non-cellulolytic bacteria can aid cellulolysis by removal of products of cellulose decomposition (Ng et al., 1981). In studies of a defined triculture degrading cellulose to methane it was found that methanogenesis rather than cellulolysis was rate-limiting (Laube and Martin, 1981, 1983). Lignin is probably not degradable by anaerobic bacteria although aromatic components of lignin released by chemical treatment and some other low molecular weight fractions of lignin can be degraded to methane by mixed cultures (Healy and Young, 1979; Colberg and Young, 1982; Zeikus et al., 1982; Grbić-Galić, 1983; Grbić-Galić and Young, 1985).

2.3.2 Soluble substrates

In the anaerobic digestion of soluble wastes either acetogenesis or methanogenesis from acetate is thought to be rate-limiting (Lawrence and McCarty, 1969; Kaspar and Wuhrmann, 1978a; Gujer and Zehnder, 1983). Under some conditions the rate of acetogenesis is controlled by the H_2-utilising methanogenic bacteria (Archer and Powell, 1985; Archer et al., 1987) and so methanogenesis by either the acetate- or H_2-utilising methanogens can be rate-limiting to the anaerobic digestion process. The properties of the acetogenic bacteria themselves are also important and there may be more than one kinetically distinct mechanism of methanogenesis from propionate (Heyes and Hall, 1983). Propionic and succinic acids appear to be readily interconvertible in digesters (Koch et al., 1983; Boone, 1984; Schink, 1985a). The importance of deciding precisely which step is truly rate-limiting can be overstated and a slight change in conditions can result in the slowest step changing from aceticlastic methanogenesis to acetogenesis (Archer et al., 1987). The sequence of events in the conversion of waste to methane is, in many ways, analogous to a sequence of enzymically-catalysed steps in a metabolic pathway. Metabolic control is a topic of much interest and it is recognised that the concept of a single rate-limiting step can be misleading (Kacser and Burns, 1973; Rapoport et al., 1974; Porteous, 1983). Each enzyme contributes to the control of the metabolic flux and procedures have been suggested by these authors for calculation of the levels of control exerted by individual enzymes. Such approaches may be applicable to mixed culture interactions and would facilitate prediction of the major controlling stages in anaerobic digestion. Irrespective of the rate-limiting steps under particular conditions, the production and release of CO_2 gas may be a substantial problem leading to process limitation (Finney and Evans, 1975; Hansson, 1979, 1982). Most problems in anaerobic digestion can be attributed to accumulation of acids and a fall in pH. Removal of acids is carried out by the methanogenic bacteria and it is therefore the methanogenic bacteria and their interactions to which attention should be focused in order to control the anaerobic digestion process. Methanogenesis from acetate and from other volatile acids are discussed in turn.

A number of aromatic components can be degraded to methane by mixed cultures (Senior and Balba, 1984; Sleat and Robinson, 1984; Wang *et al.*, 1984; Szewzyck *et al.*, 1985) including chlorinated aromatics which are dechlorinated prior to ring cleavage (Suflita *et al.*, 1982; Shelton and Tiedje, 1984). Some chlorinated aliphatic compounds have been shown to be transformed under methanogenic conditions with limited mineralisation (Bouwer and McCarty, 1983, 1985; Vogel and McCarty, 1985) although chlorinated methane analogues are inhibitory to methanogenesis (Prins *et al.*, 1972). Acetylene and ethylene are inhibitory to methanogenesis (Sprott *et al.*, 1982; Schink, 1985*b*) but some unsaturated hydrocarbons can be degraded by methanogenic enrichment cultures (Schink, 1985*c*).

2.3.3 *Methanogenesis from acetate*

Methanosarcina barkeri and *Methanosarcina mazei* are more versatile in the substrates which can be used for methanogenesis than the other acetate-using methanogenic species found in digesters, e.g. *Methanothrix soehngenii* (but see Patel and Sprott, 1990). The *Methanothrix* spp. use only acetic acid (Huser *et al.*, 1982; Fathepure, 1983; Patel, 1984; Zinder *et al.*, 1984) whereas *Methanosarcina* spp. can, in addition to acetate, use $H_2 + CO_2$, methanol and methylamines (Balch *et al.*, 1979; Mah and Smith, 1981). Thermophilic *Methanosarcina* spp. (Zinder and Mah, 1979; Ollivier *et al.*, 1984) and the mesophilic *Methanosarcina acetivorans*, not found in digesters (Sowers *et al.*, 1984), do not utilise $H_2 + CO_2$. According to the

Monod relation (Pirt, 1975) the rate of removal of acetic acid for a particular concentration is determined by the maximum specific growth rate of the acetate-utilising species and its K_s for acetate. Using published figures for μ_m and K_s for *Ms. barkeri* (Smith and Mah, 1978) and *Mt. soehngenii* (Huser *et al.*, 1982) Fig. 2.2 was constructed. This figure allows prediction of the outcome of competition for acetate by these two species. At acetate concentrations below 2·5 mM *Mt. soehngenii* should be favoured whereas *Ms. barkeri* is favoured at higher concentrations. Such predictions are based on the values of μ_m and K_s ascribed to those two isolates studied in the laboratory and must therefore be regarded with reservation. In practice high concentrations of acetate and low residence times in continuously-fed reactors favour *Ms. barkeri* (Smith *et al.*, 1980) although the presence of one species need not be to the exclusion of another.

Production of methane from acetic acid by pure cultures is described as aceticlastic (Mah *et al.*, 1978; Smith and Mah, 1978; Zehnder *et al.*,

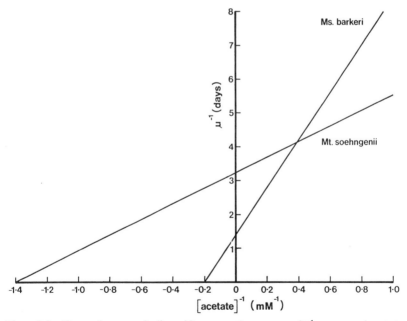

Fig. 2.2 Dependence of (specific growth rate, $\mu)^{-1}$ upon (acetate concentration)$^{-1}$ for *Methanosarcina barkeri* (Smith and Mah, 1978) and *Methanothrix soehngenii* (Huser *et al.*, 1982).

1980):

$$*CH_3^+COO^- + H_2O \rightarrow *CH_4 + H^+CO_3^- \qquad \Delta G^{o\prime} = -31 \cdot 0 \text{ kJ}$$

where * shows the conversion of the methyl group to methane.

In mixed cultures methanogenesis from acetate is also aceticlastic (Stadtman and Barker, 1949; Pine and Barker, 1956) although a mutualistic (see Table 2.2) degradation of acetate to methane is thermodynamically feasible where H_2 serves as the metabolic link between a non-methanogenic and a methanogenic bacterium:

(a) $CH_3COO^- + 4H_2O \rightarrow 2HCO_3^- + 4H_2 + H^+ \qquad \Delta G^{o\prime} = +104 \cdot 6 \text{ kJ}$

(b) $4H_2 + HCO_3^- + H^+ \rightarrow CH_4 + 3H_2O \qquad \Delta G^{o\prime} = -135 \cdot 6 \text{ kJ}$

Provided the partial pressure of H_2 is not greater than approximately 10^{-4}–10^{-5} atmospheres both reactions (a) and (b) proceed with a net negative free energy change (Fig. 2.3). A thermophilic mutualistic coculture producing methane from acetate was described by Zinder and Koch (1984). The cocultures contained a thermophilic H_2-utilising methanogenic bacterium which could not produce methane from acetate and an unidentified non-methanogenic rod-shaped bacterium. The methanogen was similar to *Methanobacterium thermoautotrophium* except that it could also utilise formate. This methanogen resembled a thermophilic rod described previously by Marty and Bianchi (1981). It has not yet proved possible to construct an acetate-utilising methanogenic mutualistic coculture from known pure species (Zinder and Koch, 1984).

Further complexity in the mechanism of methanogenesis from acetate has been revealed. In a pure culture of *Methanosarcina barkeri* H_2 was produced and consumed in large amounts during growth on CO (O'Brien *et al.*, 1984). These workers also found that during growth on either methanol or acetate, or during mixotrophic growth on CO and methanol, that H_2 was produced as a trace gas (O'Brien *et al.*, 1984). Production of H_2 was also found when *Methanosarcina* TM1 and *Methanosarcina acetivorans*, two methanogens unable to utilise $H_2 + CO_2$ as sole energy source, grew on acetate, methanol or trimethylamine (Lovley and Ferry, 1985). At the levels of H_2 commonly found in anaerobic digesters of around 10^{-4} atmospheres partial pressure (Archer, 1983) the aceticlastic methanogens are predicted to be H_2-producing organisms although the net rates of H_2 production would be very low in comparison to methane production (Lovley and Ferry, 1985). At higher concentrations of H_2 even those

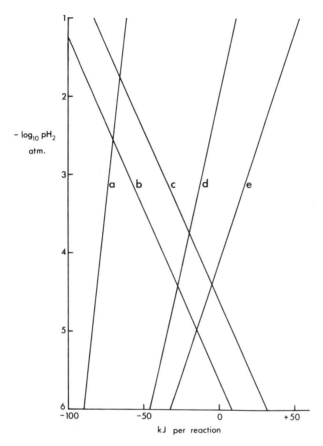

Fig. 2.3 Thermodynamic dependence of the principal reactions in anaerobic digestion upon p_{H_2}. Acetogenesis from propionate (e), butyrate (d), pyruvate (a), H_2/CO_2 (c) and methanogenesis from H_2/CO_2 (b) are shown. Calculations were based on standard values for free energies at pH 7 and 25°C with 34 mM HCO_3^-, 1 mM fatty acids and p_{CH_4} 0·7 atm. Reproduced from Archer (1983), *Enzyme and Microbial Technology*, **5**, 162–70, by permission of the publishers, Butterworth & Co. (Publishers) Ltd. ©.

aceticlastic methanogens which do not use $H_2 + CO_2$ as an energy source may become H_2-consuming bacteria due to anabolic fixation of CO_2 (Zehnder *et al.*, 1980; Lovley and Ferry, 1985). In an acetate-adapted strain of *Ms. barkeri* it was found that 14% of the 2-C was oxidixed to CO_2 with an equivalent amount of the 1-C being reduced to CH_4 (Krzycki et al., 1982) and in an acetate enrichment culture aceticlastic methanogenesis and mutualistic methanogenesis were proposed as concurrent metabolic routes (Weber et al., 1984).

2.3.4 The role of hydrogen

Hydrogen and acetate are the most important intermediary metabolites in anaerobic digestion. By definition, the levels of both H_2 and acetate do not vary substantially in concentration during steady-state anaerobic digestion. Increased H_2 levels inhibit the degradation of propionic and butyric acids due to the effect upon the thermodynamics of the reaction (Fig. 2.3) and can inhibit aceticlastic methanogenesis (Mah et al., 1978; Smith and Mah, 1978; Ferguson and Mah, 1983). Hydrogen can also inhibit methanogenesis from acetate in some methanogens which do not utilise H_2 as an energy source (Zinder and Mah, 1979; Sowers et al., 1984). This effect is not apparent in *Methanothrix* spp. (Zehnder et al., 1980; Patel, 1984). In methanogenic acetate enrichment cultures the introduction of H_2 may therefore inhibit aceticlastic methanogenesis. Acetate utilisation has been found to be inhibited by H_2 in acetate enrichment cultures (van den Berg et al., 1980) though this has not always been so (van den Berg et al., 1976). Measuring rates of methanogenesis instead of acetate utilisation as a response to H_2 in acetate enrichment cultures may not be a good guide to the effects because acetate enrichment cultures can contain obligately H_2-utilising methanogens (Archer, 1984a). Addition of H_2 to anaerobic digesters converting complex wastes to methane causes an accumulation of volatile fatty acids (Kaspar and Wuhrmann, 1978a, b; Zehnder and Koch, 1983). The possible exploitation of the role of H_2 in anaerobic digesters in monitoring and control of the anaerobic digestion process is developed in the next section.

2.4 Metabolic interactions of methanogens with other organisms

Two-species interactions have been described mathematically by imposing conditions for stability on the Lotka–Volterra equations (Lotka, 1925; Volterra, 1927). Stable interactions have been predicted in this way for predator–prey communities (Kolmogorov, 1936), and competitive and mutualistic communities (Albrecht et al., 1974;

Table 2.2 Two-species interactions

Neutralism	No interaction
Commensalism	Benefit to one species
Mutualism	Benefit to both species
Predator–prey	Benefit to one species, inhibition of other
Amensalism	Inhibition of one species
Competition	Inhibition of both species

Bulmer, 1976). It is useful here to restrict this discussion to the definitions of microbial interactions based on effects (Odum, 1953; Williamson, 1972; Bull and Slater, 1982) listed in Table 2.2.

2.4.1 Neutralism

Neutralism is difficult to prove and is of least interest, though attempting to establish that a methanogen is unaffected by the presence of another species may reveal a hitherto unknown interdependence.

2.4.2 Commensalism

There are many known examples of commensalism including those where the methanogenic species benefits and those where non-methanogenic species benefit from the presence of the methanogen. A non-specific example of commensalism is the removal of acetic acid by the aceticlastic methanogens so maintaining a neutral pH within a digester. Removal of acetic acid is beneficial to all methanogenic species which grow best at neutral pH. Another example of commensalism between different methanogenic species occurs in the rumen where *Methanobrevibacter ruminantium* requires the supply of coenzyme M for growth (Smith and Hungate, 1958). Coenzyme M is an essential cofactor in methanogenesis and is only known to be produced by methanogens (McBride and Wolfe, 1971; Balch and Wolfe, 1979). The rumen houses a number of methanogenic species which synthesise coenzyme M and, presumably, supply the needs of *Mbv. ruminantium* (Lovley *et al.*, 1984). The isolation of methanogenic bacteria on minimal media is often frustrated by the persistence of non-methanogenic bacteria even when the medium cannot support the growth in axenic culture of the non-methanogen. The growth of heterotrophic bacteria together with methanogens can be supported in media designed specifically for the growth of methanogens (Mah et al., 1977; Baresi et al., 1978; Ward et al., 1978). Non-methanogenic bacteria can depend upon the methanogen for supply of excreted nutrients and methanogens can excrete amino acids (Wellinger and Wuhrmann, 1977; Zehnder and Wuhrmann, 1977; Schönheit and Thauer, 1980) and vitamins (Ward et al., 1978). Methanogenic acetate enrichment cultures often support a highly mixed culture and the non-methanogens can then depend upon the lysis of the prime aceticlastic methanogen and use of its cellular constituents (Archer, 1984b).

The methanogenic bacteria can benefit in a commensal relationship with non-methanogenic species. Such interactions are due to the dependence of some methanogenic species upon the supply of nutrients for their growth. The nutritional requirements of the methanogenic bacteria are discussed in detail later. An example of non-specific commensalism is the removal of O_2 by the aerobic and facultatively aerobic bacteria. Some methanogens require amino acids or other factors for growth and many species are stimulated by vitamins. In the anaerobic digestion of wastes many of these nutritional requirements will not be met by the waste itself and the methanogens must depend upon other species of bacteria for their supply (Taylor, 1982).

2.4.3 Mutualism

Mutualism is a class of interaction, already briefly mentioned, which is the essential method for the methanogenic degradation of volatile fatty acids such as propionic and butyric, ethanol and other organic intermediates (Wolin, 1976, 1982; Mah, 1982). The interaction is based on the inter-species transfer of electrons (e.g. as H_2 or formate). The term was first coined for an analogous interaction between two non-methanogenic species (Iannotti et al., 1973). Methanogenic mutualistic cocultures are stable, forming mixed, or at least closely-associated, colonies on solid media (Barker, 1940; McInerney et al., 1979; Boone and Bryant, 1980) and are capable of being maintained in chemostats (Winter and Wolfe, 1980; Traore et al., 1983a, b). The importance of the mutualistic interaction in anaerobic digestion has led to an investigation of the population dynamics of mutualistic cocultures. Mathematical analysis predicts that in batch cocultures each member species has an identical specific growth rate (Powell, 1984) and that cocultures can be stably maintained in chemostats (Miura et al., 1980; Powell, 1985) provided certain qualitative conditions are satisfied.

If a coculture of two species (X and Y) converts a substrate S to product Q via an intermediate P

$$S \xrightarrow[\mu(s,p)]{X} P \xrightarrow[v(p)]{Y} Q$$

then the conditions for stability of the mutualistic interaction are:

1. specific growth rate (μ) of X increases if s increases;
2. specific growth rate (μ) of X decreases if p increases;
3. specific growth rate (v) of Y increases if p increases.

where s and p are the concentrations of S and P respectively.

Many mathematical models of the anaerobic digestion process depend upon the input of values for K_s and μ_m for the degradation of volatile fatty acids. Only as a result of having appropriate growth equations for mutualistic interactions can the true K_s and μ_m values of, say, the propionate-degrading species be determined. Apparent K_s and μ_m values for propionate degradation in particular systems can still be derived experimentally but these are composite terms of many factors (Archer et al., 1987). Experimental support for the mathematical models of mutualism (Powell, 1984, 1985; Kreikenbohm and Bohl, 1986) has been obtained for the methanogenic degradation of ethanol to methane (Fig. 2.4) in batch (Archer and Powell, 1985) and chemostat (Powell et al., 1985; Tatton et al., 1989) cocultures. The specific growth rate of a mutualistic coculture is dependent upon a number of terms but at non-limiting primary substrate concentration the specific growth rate was shown experimentally to be dependent upon the ratio of the $K_s(H_2)$ of the methanogen and its maximum

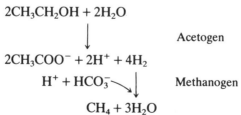

Fig. 2.4 Methanogenesis from ethanol by defined mutualistic cocultures.

specific growth rate (Archer and Powell, 1985). Thus, the properties of the methanogen determine the growth rate of the mutualistic coculture. In the complex environment of an anaerobic digester other factors such as physical proximity and metabolic interactions other than H_2 will also be important in determining the nature of interdependent mixed cultures. Interspecies transfer of formate has recently received attention (Thiele and Zeikus, 1988; Boone et al., 1989).

2.4.4 Predator–prey and amensalism

There is no evidence that methanogens are involved in predator–prey relationships. Methanogens do interact with protozoa in gastrointestinal tracts and sediments (Vogels et al., 1980; Stumm et al., 1982; Krumholz et al., 1983; van Bruggen et al., 1984) but the interaction is probably mutualistic (Odelson and Breznak, 1985). An example of an amensal reaction of methanogens is the inhibition of the growth of methanogens due to lowering of pH by the acidogenic bacteria in acidifying sludges and in the acidogenic stage of two-stage anaerobic digestion.

2.4.5 Competition

Methanogenic bacteria compete for substrates in two ways. Firstly, different methanogenic species compete among themselves and, secondly, the methanogens as a group compete with non-methanogens. Competition between methanogens for acetate has already been discussed and the outcome of competition at low acetate concentrations, is dependent, at least in part, upon relative K_s and μ_m values (Table 2.3 and Fig. 2.2). The importance of the role of K_s and μ_m must be affected by maintenance coefficient values, for which there are no data, and the ability of the methanogens to associate with other species in the formation of granules, flocs and biofilms with a digester. Competition for H_2 between methanogens, at low H_2 levels, is also determined by relative K_s and μ_m values but qualified by the considerations outlined above for aceticlastic methanogenesis. $K_s(H_2)$ values have been determined for a few methanogenic species (Kristjansson et al., 1982; Kristjansson and Schönheit, 1983; Robinson and

Table 2.3 K_s values of methanogenic and sulphate-reducing bacteria[a]

	K_s acetate (mM)	K_s H_2 (μM)
Methanosarcina barkeri	3–5	13–25
Methanothrix soehngenii	0·7	—
Methanospirillum hungatei	—	5–12
Methanobrevibacter arboriphilus	—	6
Methanobacterium thermoautotrophicum	—	8
Methanobacterium formicicum	—	6
Desulfovibrio desulfuricans	—	2
Desulfovibrio gigas	—	2
Desulfovibrio vulgaris	—	1–2
Desulfobacter postgatei	0·2	—

[a] Values from Kristjansson and Schönheit (1983) and Robinson and Tiedje (1984).

Tiedje, 1984). Only *Ms. barkeri* has a particularly high $K_s(H_2)$ value relative to other methanogens and, due also to its relatively low specific growth rate on H_2, is probably a poor competitor for H_2 in anaerobic digesters (Archer and Powell, 1985). *Ms. barkeri* must therefore depend primarily on aceticlastic methanogenesis for its growth. Maximum specific growth rates on $H_2 + CO_2$ vary widely among the methanogenic species (Taylor, 1982; Archer and Powell, 1985).

The degradation of organic matter can be diverted from methanogenesis by bacteria which utilise O_2, NO_3^- and SO_4^{2-} (Bonch-Osmolovskaya et al., 1978; Archer, 1983). The presence of sulphate in wastes presents a common problem in anaerobic digestion due to sulphate-reduction. The sulphate-reducing bacteria are effective competitors of the methanogenic bacteria for H_2 and acetate and are generally more versatile in the range of substrates utilised. The basis of the competition between sulphate-reducing bacteria and methanogens has been considered in detail elsewhere (Schönheit et al., 1982; Kristjansson and Schönheit, 1983; Robinson and Tiedje, 1984). In the presence of sulphate the sulphate-reducing bacteria (a) have a thermodynamic advantage over the methanogenic bacteria leading to generally higher μ_m values and, (b) have lower $K_s(H_2)$ and K_s(acetate) values (Table 2.3) giving them a kinetic advantage when H_2 and acetate are present at low concentrations. Sulphate-reduction in anaerobic digestion has the result that methane yields are lowered (Mosey and Hughes, 1975; Anderson et al., 1982) and the products of sulphate-reduction, especially dissolved H_2S, can be toxic to methanogenesis (Kroiss and Plahl-Wabnegg, 1983). Hydrogen-utilising and acetate-utilising sulphate-reducing bacteria are commonly found in anaerobic digesters even when digesting wastes containing low levels of sulphate (Toerien et al., 1968; Zeikus, 1980; Postgate, 1984; Richards and Archer, 1989). It must be re-emphasised, however, that the metabolic considerations discussed so far are important but are only a few of all the factors which determine competition between species. Physical properties of the respective species and proximity to H_2 or acetate-producing species will also be important in determining the outcome of competition (Lupton and Zeikus, 1984; Archer and Powell, 1985) for these methanogenic substrates.

2.5 Nutrition of the methanogenic bacteria

The nutritional requirements for the growth of methanogenic bacteria in axenic culture have been reviewed by others (Bryant et al., 1971; Mah and Smith, 1981; Taylor, 1982). This section will discuss those aspects of nutrition which are important in anaerobic digestion.

For efficient and reliable anaerobic digestion conditions for methanogenesis from $H_2 + CO_2$ and acetate must be optimised. A deficiency of an important nutrient in the waste to be digested, or an inability of the mixed culture to produce that nutrient from the waste, will have an inhibitory effect on the process. The methanogenic bacteria have a number of requirements for growth, for example, absence of oxygen and a pH generally within the range 6–8, but in many ways they are a versatile group of microorganisms because there are species which are autotrophic, thermophilic and mesophilic. Some species require salt whereas others do not (Balch et al., 1979; Mah and Smith, 1981). As a consequence most organic wastes are amenable to anaerobic digestion. A need for reliable high-rate anaerobic digestion places extra constraints on the process and then more detailed requirements of the methanogenic bacteria must be considered.

2.5.1 Trace organic nutrients

Although many methanogenic species are autotrophic the growth of most species is stimulated by the presence of vitamins (Scherer and

Sahm, 1981a) or organic supplements such as acetic acid. Other species may have requirements for acetic acid, other volatile acids or amino acids (Archer and Harris, 1985). Requirements such as these will be met in most anaerobic digesters. Propionic acid can be assimilated by some methanogenic species (Eikmanns et al., 1983) but high levels of propionic acid and other fatty acids are inhibitory to the methanogens (Hobson et al., 1974; Hobson and Shaw, 1976). Although $H_2 + CO_2$, acetate and possibly formate are the major energy-yielding substrates for methanogens in anaerobic digestion it must be remembered that other substrates can be utilised (Table 2.1) in pure culture studies. Methanol and trimethylamine have been added to wastes in order to enhance the growth of potential acetate-utilising methanogenic bacteria, and so reduce start-up times, but the efficacy of doing so has been questioned (Richards and Archer, 1989).

2.5.2 Nitrogen requirements

Methanogenic bacteria utilise the ammonium ion as a nitrogen source and in the presence of excess energy source, require 1–10 mM NH_4^+ for optimum growth (Bryant et al., 1971; Zinder and Mah, 1979; Belay et al., 1984). Amino acids may also supply nitrogen (Bhatnagar et al., 1984). It was suggested in 1954 by Pine and Barker that methanogens may be able to fix N_2 and it has been confirmed in Ms. barkeri (Murray and Zinder, 1984; Bomar et al., 1985) and Methanococcus thermolithotrophicus (Belay et al., 1984) that growth occurred in the absence of NH_4^+ by fixation of N_2. Methanogenesis by Ms. barkeri, using NH_4^+ or N_2 as the nitrogen source, is shown in Fig. 2.5. Yields of Ms. barkeri and Mc. thermolithotrophicus were reduced by a factor of 3 when N_2-grown rather than NH_4^+-grown (Belay et al., 1984; Bomar et al., 1985). Some nitrogen fixation (nif) genes from Klebsiella pneumoniae and an Anabena sp. showed homology with DNA from four different methanogenic species (Sibold et al., 1985) indicating the presence of structural genes for nitrogenase in those methanogenic species. The incidence of N_2-fixation by methanogens in anaerobic digesters has not been determined but in wastes containing low NH_4^+ levels fixation of N_2 dissolved in the waste stream may be a significant reaction. Low levels of NH_4^+ may induce a problem in buffering capacity in digesters (Anderson et al., 1982). Nitrates can be reduced to NH_4^+ and to N_2 (Gasser and Jeris, 1969; Hobson et al., 1974) in anaerobic digesters. High concentrations of NH_4^+ are inhibitory to the

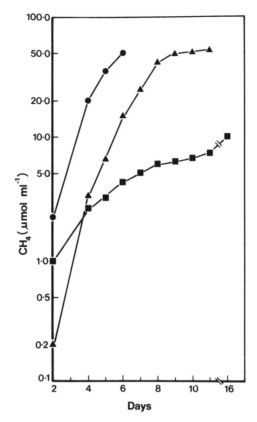

Fig. 2.5 Production of methane by *Methanosarcina barkeri* 227 under an argon headspace with 20 mM NH_4^+ (●), 14% N_2 (▲) or no nitrogen source added (■). From Murray and Zinder (1984), reprinted from *Nature*, **312**, 284–6. Copyright © 1984 Macmillan Journals Limited.

anaerobic digestion process. This inhibition, which is pH dependent and thought to be due to free ammonia, can be a problem in the digestion of some agricultural wastes (Hobson *et al.*, 1974) (see Chapter 3) and wastes from the cheese and distilling industries (Anderson *et al.*, 1982).

2.5.3 Phosphate requirements

Ms. barkeri contains approximately 40% (w/w) carbon and 1·2–1·9% (w/w) phosphorus (Scherer *et al.*, 1983). From published yield values

(Taylor, 1982; Daniels et al., 1984) it was calculated (Archer, 1985) that up to 22% of the carbon utilised by Ms. barkeri is for cell synthesis and that, for growth and methanogenesis, a C:P ratio of 100-150 is necessary. Phosphate-limited growth has been studied in Mb. thermoautotrophicum (Seely and Fahrney, 1984) and Ms. barkeri (Archer, 1985). It has been shown previously that methanogenesis by a batch-fed methanogenic calcium acetate enrichment culture showed a dramatic transition from exponentially-increasing to constant rate (Powell et al., 1983; Kirsop et al., 1984). The simplest explanation for such kinetics was that during the exponentially-increasing period methanogenesis was coupled to cell growth but was uncoupled in the constant period. In a pure culture of Ms. barkeri the kinetics were reproduced and shown to be due to phosphate limitation (Fig. 2.6). The implications of being able to control the coupling of methanogenesis to cell growth is discussed later.

2.5.4 Sulphur requirements

Methanogenic bacteria require sulphide for growth (Mountfort and Asher, 1979; Rönnow and Gunnarsson, 1981; Scherer and Sahm, 1981b) although the S requirement may sometimes be satisfied by sulphur, sulphite, thiosulphate, sulphate or amino acids (Taylor, 1982; Bhatnagar et al., 1984; Daniels et al., 1986). In excess, the products of sulphate-reduction are inhibitory to methanogenesis but some sulphate-reduction can stimulate anaerobic digestion (Khan and Trottier, 1978; van den Berg et al., 1980). The growth of Ms. barkeri in pure culture was optimal with the addition of 1–3 mM Na_2S (Mountfort and Asher, 1979; Scherer and Sahm, 1981b; Archer and King, 1984). The concentration of soluble sulphides in a digester is dependent on many factors, particularly pH and the presence of certain cations in the waste.

2.5.5 The effects of metals

Methanogenic bacteria require many metal ions for growth including sodium, potassium and magnesium (Perski et al., 1981; Sprott and Jarrell, 1981), nickel, cobalt and molybdenum (Schönheit et al., 1980), iron (Patel et al., 1978), selenium and tungsten (Jones and Stadtman,

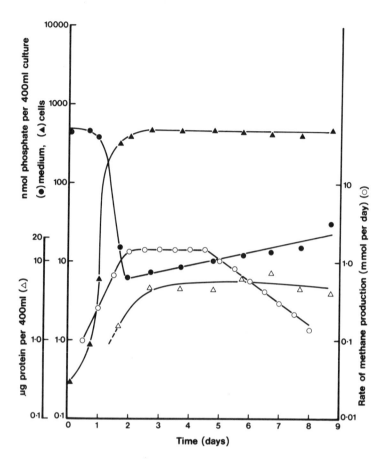

Fig. 2.6 Incorporation of ^{32}P as $KH_2\,^{32}PO_4$ from medium supernatant (●) to *Methanosarcina barkeri* strain Fusaro (▲) in 400 ml culture initially containing 1·25 μM phosphate. Rate of methanogenesis (○) and cell concentration (△) are also shown. From Archer (1985), reproduced by permission of the American Society for Microbiology.

1977, 1981). Anaerobic digestion can be stimulated by the addition of metal ions, particularly of iron and nickel (Hoban and van den Berg, 1979; Murray and van den Berg, 1981; Speece *et al.*, 1983). In wastes containing high levels of metals such as zinc, copper and chromium, toxicity can be avoided by precipitation of the insoluble sulphides (Lawrence and McCarty, 1965; Mosey and Hughes, 1975; Ashley *et al.*, 1982). In wastes such as the high-sulphate molasses effluents, however, it is the production of sulphides and consequent precipitation of essential metal ions which constitutes the problem (Richards and Archer, 1989). In some cases addition of iron salts has been successful in limiting the sulphide toxicity (Braun and Huss, 1982; Frostell, 1982) and this effect is thought to be brought about by increasing availability of iron for methanogenesis rather than a significant reduction in the H_2S concentration (Callander and Barford, 1983a, b).

2.6 Waste treatment by anaerobic mixed cultures

There are several problems and requirements associated with anaerobic digestion of wastes. Some of these difficulties can be overcome by a detailed knowledge of the biochemical and microbiological stages of the digestion process. Anaerobic digestion has a poor record of reliability whether it be with landfills, agricultural digesters or industrial waste digesters and yet a prime requirement for any company considering investing substantial sums of money in an anaerobic digester is one of reliability so that a financial return can be assured and production not put at risk by a failure of waste treatment. In landfills the methane which is produced is only a valuable energy resource if the expensive gas abstraction equipment can expect to be of useful service for 10–20 years. The digestion of agricultural wastes depends for its financial success largely on the value of the methane produced. The anaerobic digestion of industrial, usually liquid, wastes is economically attractive because effluent charges can be reduced dramatically and the methane produced is a potential on-site energy resource. Reliability in the treatment of agricultural and industrial wastes can be bought by 'over-engineering' the digester, that is, a reactor is built to a simple design with a very low waste-treatment rate. Such an approach is often not practical because of insufficient space in which to construct the reactor and because large reactors are expensive and payback times on the investment become prohibitively long. In the treatment of dilute liquid industrial wastes there is a need for reliable high rate treatment. A number of engineering configurations can achieve this but all require careful monitoring and control to ensure reliability.

It is instructive to consider what treatment rates are attainable and most information is available for the treatment of soluble wastes in single-stage reactors. An estimate of the mesophilic waste treatment rate which can be expected is approximately 1 kg COD/kg VSS. d (Henze and Harremoës, 1983). For high treatment rates per unit volume it is necessary to have high densities of biomass and a number of engineering designs aim to maintain high packing densities of bacteria. Reactor configurations are described in detail in Chapter 5 of this book and in other review articles (Ghosh and Klass, 1978; Meynell, 1982; Callander and Barford, 1983c; Fielden, 1983; Switzenbaum, 1983; Lettinga et al., 1984). In continuously-stirred reactors the bacteria and liquid waste have identical retention times and rates of treatment are then limited by the slow growth rates of the methanogenic bacteria and are always low. Bacteria can be separated from the digester effluent and returned to the reactor in the Contact Process which leads to improved treatment rates. Bacteria can be immobilised within a reactor by attachment to a support material. Then much higher treatment rates are attainable and this design is the basis of the Anaerobic Filter, Expanded Bed and Fluidised Bed Reactors. With some wastes the bacteria form a granular sludge which is dense and retained within a reactor even at short hydraulic retention times. This is the basis of the Upflow Anaerobic Sludge Blanket (UASB) Reactor. Ultrastructural studies of biofilms (Harvey et al., 1984; Richards and Turner, 1984; Robinson et al., 1984) and granules (Bochem et al., 1982; Hulshoff-Pol et al., 1983; Lettinga et al., 1988) have revealed the complexity of these structures and indicate that physical factors play an important part in the development of methanogenic mixed cultures. Comparative performance figures of the aceticlastic methanogenic bacteria and various single-stage mesophilic anaerobic digesters are given in Table 2.4. The treatment rates given for the pure cultures of aceticlastic methanogens are maximum values achieved during the exponential phase of growth in fed-batch cultures. Those figures are therefore a guide to what is achievable but are unlikely to be approached in an anaerobic digester. Treatment rates given for the anaerobic digesters show the typical range of values published which are heavily dependent on the waste composition and operating conditions. Values outside the ranges given can be found but most digesters will operate within the given ranges. Extremely high biomass concentrations have been reported in attached film (Jewell, 1980) and UASB (Lettinga et al., 1980) reactors demonstrating the potential of these two types of reactor for the rapid treatment of wastes.

Table 2.4 Performance figures for mesophilic methanogenesis

	Methanosarcina barkeri	Methanothrix soehngenii	Methanothrix concilii	Contact reactor	Attached film	UASB[a] reactor
Reference	1, 2	3	4	5	5, 6	5, 7
Biomass concentration (kg VSS/m^3)	0·02–0·05[b]	0·03[b]	0·004[b]	5–20	10–40	15–50
COD removal (kg COD/m^3 day)	0·35	0·30	0·20	1–8	5–20	5–30
Methane yield[c]						
(mol CH$_4$/m^3 day)	5	4·3	3·0	15–120	75–300	75–450
(m^3 CH$_4$/m^3 day)	0·12 [d]	0·10 [d]	0·07 [d]	0·3–2·7	1·5–7	1·5–10·5
Specific rates						
(kg COD/kg VSS day)	7–21	9–12	55	0·2–1·0	0·5–1·0	0·5–1·0
(mol CH$_4$/kg VSS day)	110–330	140–180	870	3–15	7·5–15	7·5–15

[a] Assume VSS ≃ 75% TS.
[b] During exponential growth of fed-batch cultures.
[c] Assuming 1 kg COD → 15 mol CH$_4$.
[d] From metabolic quotient $q_m = \mu m/y$.

References: 1. Smith and Mah (1978).
2. Mah et al. (1978).
3. Huser et al. (1982).
4. Patel (1984).
5. Callander and Barford (1983c).
6. Henze and Harremoës (1983).
7. Lettinga et al. (1984).

2.6.1 Monitoring anaerobic digestion

The information given in Table 2.4 illustrates the value in measuring parameters such as COD removal, VSS and CH_4 yields in assessing the operation of a digester. The discussion in this chapter has shown that the methanogenic bacteria hold the key to successful digester operation and that their concentration and metabolic rates are useful monitoring parameters. pH is an important factor to be measured directly and in relation to the volatile fatty acid concentrations, alkalinity and CO_2 levels. The concentration of H_2 determines the rate of degradation of some intermediate compounds by mutualistic cocultures and is a potentially useful monitoring parameter in process control. All of these parameters can be measured and their value depends upon the interpretation of the data in microbiological terms (Archer, 1983). Methodologies for standard assays such as pH, BOD, COD, VSS, TS, CH_4, CO_2, alkalinity, VFA levels can be found elsewhere (Standard Methods, 1960; Meynell, 1982; Rozzi and Labellarte, 1984). These data can be used to check whether the specific rates of COD removal are as expected and that the pH is approximately neutral allowing efficient methanogenesis. pH probes can operate in digesters despite being prone to fouling (Kennedy and Muzar, 1984). In the case of an imbalance between acidogenesis and methanogenesis a neutral pH may be maintained at the expense of buffering capacity which would lead ultimately to acidification. Alkalinity measurements are therefore useful in conjunction with pH measurements and can be exploited in microprocessor-based control (Rozzi and Labellarte, 1984; Wheatley *et al.*, 1987).

Measurement of the total biomass (by VSS or TS) is usually adequate in digester monitoring but detailed information on the content of the methanogenic bacteria may be required in order to ensure that the specific rate of metabolism is not impaired. Direct isolation and cultivation of the methanogenic species is slow and often unreliable although purification of a species need not be required to estimate the aceticlastic methanogenic capacity of a digester; activity tests give an indication of the ability of the test system to produce methane from a particular substrate (Koepp *et al.*, 1983; Valcke and Verstraete, 1983; Dolfing and Bloemen, 1985). Serological probing of mixed cultures for methanogenic bacteria is of potential value in qualitative and quantitative studies (Strayer and Tiedje, 1978; Archer, 1984*c*). Use of monoclonal antibodies has demonstrated the applicability of serological methodologies in probing anaerobic digesters (Kemp et al., 1986). The methanogenic bacteria are sufficiently different to the eubacteria for there to be chemically-based assays offering specificity. Assay of polyether lipids has been used to estimate the amount of methanogenic biomass in environmental samples (Martz et al., 1983) and this methodology could be applied to anaerobic digesters. Coenzyme F_{420} is found in all methanogenic bacteria (Eirich et al., 1979) but less commonly and in lower concentrations in non-methanogenic bacteria (Eker et al., 1980; Vogels et al., 1984; Daniels et al., 1985). Assay of coenzyme F_{420} by high-performance liquid chromatography and fluorescence detection can give an estimate of the methanogenic biomass in digesters (van Beelen et al. 1983a, b). There are structural differences in coenzyme F_{420} molecules from different species, and this technique has potential for estimating the amounts of different methanogenic species. Fluorimetric monitoring of extracts of digesters without separation of extracted compounds by HPLC (Delafontaine et al., 1979) can give an unreliable estimate of the amount of coenzyme F_{420} and methanogenic bacteria (Archer, 1984b; Dolfing and Mulder, 1985; Zabranska et al., 1985). Methods for assaying methanogenic bacteria have recently been reviewed (Peck and Archer, 1989).

2.6.2 Control of nutrients

The nutritional requirements of the methanogenic bacteria have been described in detail. A waste must supply the nutritional needs of the bacteria for successful anaerobic digestion. Examples have been given earlier when the micronutrients were found to be important but, while most wastes will be sufficient in the minor elements, it is important to ensure that the COD:N ratio is suitable (van den Berg and Lenz, 1981; Anderson et al., 1982; Meynell, 1982). Often overlooked is the importance of phosphate. It has already been described how phosphate can control the coupling between methanogenesis and cell growth. During start-up of a digester it is essential to encourage coupling, i.e. encourage biomass formation. In the operation of a fully-commissioned digester, particularly when equipped with biomass retention, production of excess biomass can be an undesirable embarrassment. Phosphate-limitation allows high-rate methanogenesis without excess biomass formation. Control over the coupling of methanogenesis and cell growth would be advantageous in digester operation and is feasible, although as yet untried, in the anaerobic digestion of naturally phosphate-deficient wastes. The waste would receive supplementary phosphate during start-up but not during steady-state operation. Suitable phosphate-deficient wastes are landfill leachates (Robinson *et al.*, 1982; Bull *et al.*, 1983), some starch wastes (Lettinga *et al.*, 1983) and some molasses-based effluents (Sanchez Riera *et al.*, 1982; Szendrey, 1983).

2.6.3 Monitoring hydrogen

The concentration of H_2 has been known for some time to exert a controlling influence on the degradation of a number of organic compounds which are important intermediates in anaerobic digestion (Hungate, 1975; Wolin, 1976, 1982). In an extension of this work the concentration of H_2 may also exert a controlling influence over the rate of anaerobic digestion (Mosey, 1982). A rise in the levels of H_2 would be expected to inhibit the further degradation of volatile fatty acids, including acetic, leading to a lowering of pH in the digester. Low pH selectively inhibits the methanogenic bacteria resulting in a further and catastrophic acidification within the digester as volatile fatty acids accumulate. Two-stage reactors, where acidogenesis and methanogenesis occur in separate compartments, have an advantage here because the pH of effluent from the acidogenic stage can be easily regulated prior to introduction into the methanogenic reactor (Ghosh and Klass, 1978; Cohen *et al.*, 1982). A rise in levels of H_2 is therefore a potentially very early indicator of acidification in a digester (Gujer and Zehnder, 1983; Mosey and Fernandes, 1984; Archer *et al.*, 1987). Addition of H_2 to digesters leads to a rise in propionic acid levels (Kaspar & Wuhrmann, 1978*a, b*) although anaerobic digesters generally have a large reserve capacity for H_2-utilisation because steady-state levels of H_2 are low enough to be below the $K_s(H_2)$ values of the methanogenic bacteria (Archer, 1983; Zinder, 1984).

Measurement of H_2 in the biogas from a digester has the advantage that the measuring system, if on-line, is less prone to fouling by the digester contents but the disadvantage that the H_2 in the gas may not be in equilibrium with the H_2 in solution. Low levels of H_2 in gas can be measured using gas chromatography (Robinson and Tiedje, 1982, 1984; Schönheit *et al.*, 1980). Detection of H_2 in digester gas by a polarographic method (Bartlett *et al.*, 1980) has given some experimental indication that H_2 is a useful parameter to be measured in control of laboratory and industrial-scale digesters (Mosey and Fernandes, 1984; Archer *et al.*, 1986). Measurement of dissolved gases by membrane inlet mass spectrometry has proved to be a very useful research tool for monitoring H_2 in digesters (Scott *et al.*, 1983; Whitmore *et al.*, 1985) and initiating feed-back control (Whitmore and Lloyd, 1986). Addition of glucose, propionate or butyrate to mesophilic digesters caused a rapid rise in the dissolved H_2 levels (Whitmore *et al.*, 1985). Development of this methodology could lead to more detailed studies of the effects of shock loadings on digesters and the development of remedial measures, based on measurement of H_2, to cope with acidification. Certainly it should be possible to arrange that changed hydrogen levels can be translated into physical action, for example by initiating the monitoring of buffer capacity on-line (Powell and Archer, 1989). Indeed, the monitoring and control of anaerobic digesters depends upon the development of appropriate sensors for all the parameters listed above. Sensors are being developed for many of the most useful parameters (Lowe, 1985) including biomass (Harris and Kell, 1985). Specific sensors for individual bacterial species based upon serology (Harris and Kell, 1985) also seems possible. The need for reliability of all sensory equipment in the complex environment of an anaerobic digester places extra constraints on the development of this technology.

2.7 Conclusions

Complex mixtures of bacteria are required to convert organic wastes to methane in the anaerobic digestion process. The microbiology of the process is becoming better understood in forms of improved knowledge of individual species present and in relation to the basis of the interactions between species. Applications of this knowledge will lead to improvements in anaerobic digestion because the process will then become amenable to control. Improvements in the process are required to achieve reliable waste treatment, preferably at high rates, to make the process more cost effective.

Efficient waste treatment by anaerobic digestion depends crucially on the methanogenic species. Two functional groups of methanogenic species are recognised, those which remove hydrogen and those which remove acetic acid. The stability of the anaerobic digestion process relies upon interactions between the methanogens and other species of bacteria. Monitoring procedures, based for example on detection of levels of hydrogen in the digester gas, are being designed to provide a sensitive and early warning of instability between methanogens and their interacting species. Appropriate action can then be triggered to return the mixed culture to stability before an irreversible acidification of the digester occurs causing failure. Effective control requires a deterministic model of the process which, in turn, can only be developed from an improved understanding of the microbial interactions.

2.8 References

Adamse, A. D. (1980). New isolation of *Clostridium aceticum* (Wieringa). *Antonie van Leeuwenhoek*, **46**, 523–31.
Albrecht, F., Gatzke, H., Haddad, A. and Wax, N. (1974). The dynamics of two interacting populations. *Journal of Mathematical and Analytical Applications*, **46**, 658–70.
Anderson, G. K., Donnelly, T. and McKeown, K. J. (1982). Identification and control of inhibition in the anaerobic treatment of industrial wastewaters. *Process Biochemistry*, **17**(4), 28–32, 41.
Archer, D. B. (1983). The microbiological basis of process control in methanogenic fermentation of soluble wastes. *Enzyme and Microbial Technology*, **5**, 162–70.
Archer, D. B. (1984a). Biochemistry of methanogenesis by mixed cultures. *Biochemistry Society Transactions*, **12**, 1144–6.
Archer, D. B. (1984b). Hydrogen-using bacteria in a methanogenic acetate enrichment culture. *Journal of Applied Bacteriology*, **56**, 125–9.
Archer, D. B. (1984c). Detection and quantitation of methanogens using enzyme-linked immunosorbent assay. *Applied and Environmental Microbiology*, **48**, 797–801.
Archer, D. B. (1985). Uncoupling of methanogenesis from growth of *Methanosarcina barkeri* by phosphate limitation. *Applied and Environmental Microbiology*, **50**, 1233–7.
Archer, D. B. and Harris, J. E. (1985). Methanogenic bacteria and methane production in various habitats. In *Anaerobic Bacteria in Habitats other than Man*, ed. G. C. Mead and E. M. Barnes. Blackwell Scientific Publications, Oxford, pp. 185–223.
Archer, D. B. and King, N. R. (1984). Isolation of gas vesicles from *Methanosarcina barkeri*. *Journal of General Microbiology*, **130**, 167–72.
Archer, D. B. and Powell, G. E. (1985). Dependence of the specific growth rate of methanogenic mutualistic cocultures on the methanogen. *Archives of Microbiology*, **141**, 133–7.

Archer, D. B., Hilton, M. G., Adams, P. and Wiecko, H. (1986). Hydrogen as a process control index in a pilot scale anaerobic digester. *Biotechnology Letters*, **8**, 197–202.
Archer, D. B., Powell, G. E., Hilton, M. G. and Tatton, M. J. (1987). Control of anaerobic digestion by H_2- and acetate-utilising methanogenic bacteria. In *Bioenvironmental Systems II*, ed. D. L. Wise. CRC Press Inc, Boca Raton, Florida, pp. 113–34.
Ashley, N. V., Davies, M. and Hurst, T. J. (1982). The effect of increased nickel ion concentrations on microbial populations in the anaerobic digestion of sewage sludge. *Water Research*, **16**, 963–71.
Balch, W. E. and Wolfe, R. S. (1979). Specificity and biological distribution of coenzyme M (2-mercaptoethanesulfonic acid). *Journal of Bacteriology*, **137**, 256–63.
Balch, W. E., Schoberth, S., Tanner, R. S. and Wolfe, R. S. (1977). Acetobacterium, a new genus of hydrogen-oxidising, carbon dioxide-reducing, anaerobic bacteria. *International Journal of Systematic Bacteriology*, **27**, 355–61.
Balch, W. E., Fox, G. E., Magrum, L. J., Woese, C. R. and Wolfe, R. S. (1979). Methanogens: reevaluation of a unique biological group. *Microbiological Reviews*, **43**, 260–96.
Baresi, L., Mah, R. A., Ward, D. M. and Kaplan, I. R. (1978). Methanogenesis from acetate: enrichment studies. *Applied and Environmental Microbiology*, **36**, 186–97.
Barker, H. A. (1940). Studies upon the methane fermentation IV. The isolation and culture of *Methanobacterium omelianskii*. *Antonie van Leeuwenhoek*, **6**, 201–20.
Barker, H. A. (1956). *Bacterial Fermentation*. Wiley, New York.
Bartlett, K., Dobson, J. V. and Eastham, E. (1980). A new method for the detection of hydrogen in breath and its application to acquired and inborn sugar malabsorption. *Clinica Chimica Acta*, **108**, 189–94.
Belay, N., Sparling, R. and Daniels, L. (1984). Dinitrogen fixation by a thermophilic methanogenic bacterium. *Nature*, **312**, 286–8.
Bhatnagar, L., Jain, M. K., Aubert, J.-P. and Zeikus, J. G. (1984). Comparison of assimilatory organic, nitrogen, sulphur, and carbon dioxide sources for growth of *Methanobacterium* species. *Applied and Environmental Microbiology*, **48**, 785–90.
Bochem, H. P., Schoberth, S. M., Sprey, B. and Wengler, P. (1982). Thermophilic biomethanation of acetic acid: morphology and ultrastructure of a granular consortium. *Canadian Journal of Microbiology*, **28**, 500–10.
Bomar, M., Knoll, K. and Widdel, F. (1985). Fixation of molecular nitrogen by *Methanosarcina barkeri*. *FEMS Microbiology Ecology*, **31**, 47–55.
Bonch-Osmolovskaya, E. A., Vedenino, I. Y. and Balashova, V. V. (1978). Influence of inorganic electron acceptors on the bacterial formation of methane from cellulose. *Microbiology* (Eng. translation of *Mikrobiologiya*), **47**, 487–2.
Boone, D. R. (1982). Terminal reactions in the anaerobic digestion of animal waste. *Applied and Environmental Microbiology*, **43**, 57–64.

Boone, D. R. (1984). Propionate exchange reactions in methanogenic ecosystems. *Applied and Environmental Microbiology*, **48**, 863–4.
Boone, D. R. and Bryant, M. P. (1980). Propionate-degrading bacterium, *Syntrophobacter wolinii* sp. nov., gen. nov., from methanogenic ecosystems. *Applied and Environmental Microbiology*, **40**, 626–32.
Boone, D. R. and Whitman, W. B. (1988). Proposals of minimal standards for describing new taxa of methanogenic bacteria. *International Journal of Systematic Bacteriology*, **38**, 212–19.
Boone, D. R., Johnson, R. L. and Liu, Y. (1989). Diffusion of the interspecies electron carriers H_2 and formate in methanogenic ecosystems and its implication in the measurement of K_m for H_2 or formate uptake. *Applied and Environmental Microbiology*, **55**, 1735–41.
Bouwer, E. J. and McCarty, P. L. (1983). Transformation of 1- and 2-carbon halogenated aliphatic organic compounds under methanogenic conditions. *Environmental and Scientific Technology*, **45**, 1286–94.
Bouwer, E. J. and McCarty, P. L. (1985). Ethylene dibromide transformation under methanogenic conditions. *Applied and Environmental Microbiology*, **50**, 527–8.
Braun, M. and Gottschalk, G. (1982). *Acetobacterium wieringae* sp. nov., a new species producing acetic acid from molecular hydrogen and carbon dioxide. *Zentralblatt für Bacteriologie und Hygiene Abt. 1, Orig. C*, **3**, 368–76.
Braun, M., Schoberth, S. and Gottschalk, G. (1979). Enumeration of bacteria forming acetate from H_2 and CO_2 in anaerobic habitats. *Archives of Microbiology*, **120**, 201–4.
Braun, R. and Huss, S. (1982). Anaerobic digestion of distillery effluents. *Process Biochemistry*, **17**(4), 25–7.
Bryant, M. P. (1979). Microbial methane production—theoretical aspects. *Journal of Animal Science*, **48**, 193–201.
Bryant, M. P., Wolin, E. A., Wolin, M. J. and Wolfe, R. S. (1967). *Methanobacillus omelianskii*, a symbiotic association of two species of bacteria. *Archiv für Mikrobiologie*, **59**, 20–31.
Bryant, M. P., Tzeng, S. F., Robinson, I. M. and Joyner, A. E. (1971). Nutrient requirements of the methanogenic bacteria. In *Anaerobic Biological Treatment Processes*, ed. F. G. Pohland. *Advances in Chemistry Series*, **105**. American Chemical Society, Washington, DC, pp. 23–40.
Bryant, M. P., Campbell, L. L., Reddy, C. A. and Crabill, M. R. (1977). Growth of *Desulfovibrio* in lactate or ethanol media low in sulphate in association with H_2-utilising methanogenic bacteria. *Applied and Environmental Microbiology*, **33**, 1162–9.
Bull, A. T. and Slater, J. H. (1982) (eds). Microbial interactions and community structure. In *Microbial Interactions and Communities*, Vol. 1. Academic Press, London, pp. 13–44.
Bull, P. S., Evans, J. V., Wechsler, R. M. and Cleland, K. J. (1983). Biological technology of the treatment of leachate from sanitary landfills. *Water Research*, **17**, 1473–81.
Bulmer, M. J. (1976). The theory of prey–predator oscillation. *Theoretical Population Biology*, **9**, 137–50.

Callander, I. J. and Barford, J. P. (1983a). Precipitation, chelation, and the availability of metals as nutrients in anaerobic digestion. 1. Methodology. *Biotechnology and Bioengineering*, **25**, 1947–57.

Callander, I. J. and Barford, J. P. (1983b). Precipitation, chelation, and the availability of metals as nutrients in anaerobic digestion. II. Applications. *Biotechnology and Bioengineering*, **25**, 1959–72.

Callander, I. J. and Barford, J. P. (1983c). Recent advances in anaerobic digestion technology. *Process Biochemistry*, **18**(4), 24–30, 37.

Chynoweth, D. P. and Mah, R. A. (1971). Volatile acid formation in sludge digestion. In *Anaerobic Biological Treatment Processes*, ed. F. G. Pohland. *Advances in Chemistry Series*, **105**. American Chemical Society, Washington, DC, pp. 41–54.

Cohen, A., Breure, A. M., van Andel, J. G. and Deursen, A. (1982). Influence of phase separation on the anaerobic digestion of glucose. II. Stability and kinetic responses to shock loadings. *Water Research*, **16**, 449–55.

Colberg, P. J. and Young, L. Y. (1982). Biodegradation of lignin-derived molecules under anaerobic conditions. *Canadian Journal of Microbiology*, **28**, 886–9.

Cromwell, D. L. (1965). Identification of microflora present in sanitary landfills. MSc Thesis. West Virginia University.

Daniels, L., Sparling, R. and Sprott, G. D. (1984). The bioenergetics of methanogenesis. *Biochimica et Biophysica Acta*, **768**, 113–63.

Daniels, L., Bakhiet, N. and Harmon, K. (1985). Widespread distribution of a 5-deazaflavin cofactor in actinomyces and related bacteria. *Systematic and Applied Microbiology*, **6**, 12–17.

Daniels, L., Belay, N. and Rajagopal, B. S. (1986). Assimilatory reduction of sulfate and sulfite by methanogenic bacteria. *Applied and Environmental Microbiology*, **51**, 703–9.

Delafontaine, M. J., Naveau, H. P. and Nyns, E. J. (1979). Fluorimetric monitoring of methanogenesis in anaerobic digesters. *Biotechnology Letters*, **1**, 71–4.

Dolfing, J. and Bloemen, W. G. B. M. (1985). Activity measurements as a tool to characterize the microbial composition of methanogenic environments. *Journal of Microbiological Methods*, **4**, 1–12.

Dolfing, J. and Mulder, J.-W. (1985). Comparison of methane production rate and coenzyme F420 content of methanogenic consortia in anaerobic granular sludge. *Applied and Environmental Microbiology*, **49**, 1142–5.

Dunnill, P. and Rudd, M. (1984). *Biotechnology and British Industry*. Science and Engineering Research Council, London.

Eastman, J. A. and Ferguson, J. F. (1981). Solubilization of particulate organic carbon during the acid phase of anaerobic digestion. *Journal of Water Pollution Control Federation*, **53**, 352–66.

Eichler, B. and Schink, B. (1984). Oxidation of primary aliphatic alcohols by *Acetobacterium carbinolicum* sp. nov., a homoacetogenic anaerobe. *Archives of Microbiology*, **140**, 147–52.

Eikmanns, B., Jaenchen, R. and Thauer, R. K. (1983). Propionate assimilation by methanogenic bacteria. *Archives of Microbiology*, **136**, 106–10.

Eirich, L., Vogels, G. D. and Wolfe, R. S. (1979). Distribution of coenzyme

F420 and properties of its hydrolytic fragments. *Journal of Bacteriology,* **140**, 20–7.

Eker, A. P. M., Pol, A., van der Meijden, P. and Vogels, G. D. (1980). Purification and properties of 8-hydroxy-5-deazaflavin derivatives from *Streptomyces griseus*. *FEMS Microbiology Letters,* **8**, 161–5.

Fathepure, B. Z. (1983). Isolation and characterisation of an aceticlastic methanogen from a biogas digester. *FEMS Microbiology Letters,* **19**, 151–6.

Ferguson, T. J. and Mah, R. A. (1983). Effect of H_2–CO_2 on methanogenesis from acetate or methanol in *Methanosarcina* spp. *Applied and Environmental Microbiology,* **46**, 348–55.

Fielden, N. E. H. (1983). The theory and practice of anaerobic digestion reactor design. *Process Biochemistry,* **18**(5), 34–7.

Filip, Z. and Küster, E. (1979). Microbial activity and the turnover of organic matter is municipal refuse disposed of in a landfill. *European Journal of Applied Microbiology and Biotechnology,* **7**, 371–9.

Finney, C. D. and Evans, R. S. (1975). Anaerobic digestion: the rate-limiting process and the nature of inhibition. *Science,* **190**, 1088–9.

Frostell, B. (1982). Anaerobic fluidized bed experimentation with a molasses waste water. *Process Biochemistry,* **17**, 37–40.

Gasser, R. and Jeris, J. S. (1969). Comparison of various nitrogen sources in anaerobic treatment. *Journal of Water Pollution Control Federation,* **41**, R91–R100.

Ghosh, S. and Klass, D. L. (1978). Two-phase anaerobic digestion. *Process Biochemistry,* **13**(4), 15–24.

Grbić-Galić, D. (1983). Anaerobic degradation of coniferyl alcohol by methanogenic consortia. *Applied and Environmental Microbiology,* **46**, 1442–6.

Grbić-Galić, D. and Young, L. J. (1985). Methane fermentation of ferulate and benzoate: anaerobic degradation pathways. *Applied and Environmental Microbiology,* **50**, 292–7.

Gujer, W. and Zehnder, A. J. B. (1983). Conversion processes in anaerobic digestion. *Water Science and Technology,* **15**, 127–67.

Hansson, G. (1979). Effects of carbon dioxide and methane on methanogens. *European Journal of Applied Microbiology and Biotechnology,* **6**, 351–9.

Hansson, G. (1982). End product inhibition in methane fermentation. *Process Biochemistry,* **17**(6), 45–9.

Harmsen, J. (1983). Identification of organic compounds in leachate from a waste tip. *Water Research,* **17**, 699–705.

Harris, C. M. and Kell, D. B. (1985). The estimation of microbial biomass. *Biosensors,* **1**, 17–84.

Harvey, M., Forsberg, C. W., Beveridge, T. J., Pos, J. and Ogilvie, J. R. (1984). Methanogenic activity and structural characteristics of the microbial biofilm on a needle-punched polyester support. *Applied and Environmental Microbiology,* **48**, 633–8.

Healy, J. B. & Young, C. Y. (1979). Aromatic degradation of eleven aromatic compounds to methane. *Applied and Environmental Microbiology,* **38**, 84–9.

Henson, J. M. and Smith, P. H. (1985). Isolation of a butyrate-utilizing bacterium in coculture with *Methanobacterium thermoautotrophicum* from a thermopholic digester. *Applied and Environmental Microbiology*, **49**, 1461–6.
Henze, M. and Harremoës, P. (1983). Anaerobic treatment of wastewater in fixed film reactors—a literature review. *Water Science and Technology*, **15**, 1–101.
Heukelekian, H. and Mueller, P. (1958). Transformation of some lipids in anaerobic sludge digestion. *Sewage and Industrial Wastes*, **30**, 1108–20.
Heyes, R. H. and Hall, R. J. (1983). Kinetics of two subgroups of propionate-using organisms in anaerobic digestion. *Applied and Environmental Microbiology*, **46**, 710–15.
Hoban, D. J. and van den Berg, L. (1979). Effect of iron on conversion of acetic acid to methane during methanogenic fermentations. *Journal of Applied Bacteriology*, **47**, 153–9.
Hobson, P. N. (1981). Microbial pathways and interactions in the anaerobic treatment process. In *Mixed Culture Fermentations*, ed. M. E. Bushell and J. H. Slater. SGM Special Publication 5, Academic Press, London, pp. 53–79.
Hobson, P. N. and Shaw, B. G. (1976). Inhibition of methane production by *Methanobacterium formicicum*. *Water Research*, **10**, 849–52.
Hobson, P. N., Bousfield, S. and Summer, R. (1974). Anaerobic digestion of organic matter. *CRC Critical Reviews in Environmental Control*, **4**, 131–91.
Hulshoff-Pol, L. W., de Zeeuw, W. J., Velzeboer, C. T. M. and Lettinga, G. (1983). Granulation in UASB reactors. *Water Science and Technology*, **15**, 291–304.
Hungate, R. E. (1975). The rumen microbial ecosystem. *Annual Review of Ecology and Systematics*, **6**, 39–66.
Huser, B. A., Wuhrmann, K. and Zehnder, A. J. B. (1982). *Methanothrix soehngenii* gen. nov. sp. nov., a new acetotrophic non-hydrogen-oxidizing methane bacterium. *Archives of Microbiology*, **132**, 1–9.
Iannotti, E. L., Kafkewitz, D., Wolin, M. J. and Bryant, M. P. (1973). Glucose fermentation products of *Ruminococcus albus* grown in continuous culture with *Vibrio succinogenes*: changes caused by interspecies transfer of H_2. *Journal of Bacteriology*, **114**, 1231–40.
Iannotti, E. L., Fischer, J. R. and Sievers, D. M. (1982). Characterisation of bacteria from a swine manure digester. *Applied and Environmental Microbiology*, **43**, 136–43.
Jewell, W. J. (1980). Development of the attached microbial film expanded-bed process for aerobic and anaerobic waste treatment. In *Biological Fluidised Bed Treatment of Water and Wastewater*, ed. P. F. Cooper and B. Atkinson. Ellis Horwood, Chichester, pp. 251–69.
Jones, J. B. and Stadtman, T. C. (1977). *Methanococcus vannielii*: culture and effects of selenium and tungsten on growth. *Journal of Bacteriology*, **130**, 1404–6.
Jones, J. B. and Stadtman, T. C. (1981). Selenium-dependent and selenium-independent formate dehydrogenases of *Methanococcus vannielii*.

Separation of the two forms and characterisation of the purified selenium-independent form. *Journal of Biological Chemistry*, **256**, 656–63.
Jones, K. L. and Grainger, J. M. (1983). The application of enzyme activity measurements to a study of factors affecting protein, starch and cellulose fermentation in domestic refuse. *European Journal of Applied Microbiology and Biotechnology*, **18**, 181–5.
Jones, K. L., Rees, J. F. and Grainger, J. M. (1983). Methane generation and microbial activity in a domestic refuse landfill site. *European Journal of Microbiology and Biotechnology*, **18**, 242–5.
Kacser, H. and Burns, J. A. (1973). The control of flux. *Symposium of the Society for Experimental Biology*, **27**, 65–104.
Kaspar, H. F. and Wuhrmann, K. (1978a). Kinetic parameters and relative turnovers of some important catabolic reactions in digesting sludge. *Applied and Environmental Microbiology*, **36**, 1–7.
Kaspar, H. F. and Wuhrmann, K. (1978b). Product inhibition in sludge digestion. *Microbial Ecology*, **4**, 241–8.
Kemp, H. A., Morgan, M. R. A. and Archer, D. B. (1986). Enzyme-linked immunosorbent assay for methanogens using polyclonal and monoclonal antibodies. *Proceedings of Anaerobic Treatment. A Grown-Up Technology*, Amsterdam, Industrial Presentations (Europe) BV, pp. 39–50.
Kennedy, K. J. and Muzar, M. (1984). Continuous pH measurement for anaerobic digestion. *Biotechnology and Bioengineering*, **26**, 632.
Khan, A. W. (1980). Degradation of cellulose to methane by a coculture of *Acetovibrio cellulolyticus* and *Methanosarcina barkeri*. *FEMS Microbiology Letters*, **9**, 233–5.
Khan, A. W. and Trottier, T. M. (1978). Effect of sulphur-containing compounds on anaerobic degradation of cellulose to methane by mixed cultures obtained from sewage sludge. *Applied and Environmental Microbiology*, **35**, 1027–34.
Kirsop, B. H. (1984). Methanogenesis. *CRC Critical Reviews in Biotechnology*, **1**, 109–59.
Kirsop, B. H., Hilton, M. G., Powell, G. E. and Archer, D. B. (1984). Methanogenesis in the anaerobic treatment of food-processing wastes. In *Microbiological Methods for Environmental Biotechnology*, ed. J. M. Grainger and J. M. Lynch. Academic Press, London, pp. 139–58.
Koch, M., Dolfing, J., Wuhrmann, K. and Zehnder, A. J. B. (1983). Pathways of propionate degradation by enriched methanogenic cultures. *Applied and Environmental Microbiology*, **45**, 1411–14.
Koepp, H.-J., Schoberth, S. M. and Sahm, H. (1983). Evaluation of the anaerobic digestion of an effluent from citric acid fermentation. In *Anaerobic Waste Water Treatment*, ed. W. J. van den Brink. TNO Corporate Communication Department, The Hague, pp. 13–29.
Kolmogorov, A. (1936). Sulla teoria di Volterra della lotta per l'esistenza. *Giornale Instituto Italiano degli Attuari*, **7**, 74–80; translated in *Applicable Mathematics of Non-Physical Phenomena* (1982), ed. F. Oliveira-Pinto and B. W. Conolly. Ellis Horwood, Chichester, pp. 117–22.

Kreikenbohm, R. and Bohl, E. (1986). A mathematical model of syntrophic cocultures in the chemostat. *FEMS Microbiology Ecology*, **38**, 131–40.

Kristjansson, J. K. and Schönheit, P. (1983). Why do sulphate-reducing bacteria outcompete methanogenic bacteria for substrates? *Oecologia*, **60**, 264–6.

Kristjansson, J. K., Schönheit, P. and Thauer, R. K. (1982). Different K_s values for hydrogen of methanogenic bacteria and sulphate reducing bacteria: an explanation for the apparent inhibition of methanogenesis by sulphate. *Archives of Microbiology*, **131**, 278–82.

Kroiss, H. and Plahl-Wabnegg, F. (1983). Sulphide toxicity with anaerobic waste water treatment. In *Anaerobic Wastewater Treatment*, ed. W. J. van den Brink. TNO Corporate Communication Department, The Hague, pp. 72–85.

Krumholz, L. R. and Bryant, M. P. (1986). *Syntrophococcus sucromutans* sp. nov. gen. nov. uses carbohydrates as electron donors and formate, methoxymonobenzenoids or *Methanobrevibacter* as electron acceptor systems. *Archives of Microbiology*, **143**, 313–18.

Krumholz, L. R., Forsberg, C. W. and Veira, D. M. (1983). Association of methanogenic bacteria with rumen protozoa. *Canadian Journal of Microbiology*, **29**, 676–80.

Krzycki, J. A., Wolkin, R. H. and Zeikus, J. G. (1982). Comparison of unitrophic and mixotrophic substrate metabolism by an acetate-adapted strain of *Methanosarcina barkeri*. *Journal of Bacteriology*, **149**, 247–54.

Laube, V. M. and Martin, S. M. (1981). Conversion of cellulose to methane and carbon dioxide by triculture of *Acetivibrio cellulolyticus*, *Desulfovibrio* sp. and *Methanosarcina barkeri*. *Applied and Environmental Microbiology*, **42**, 413–20.

Laube, V. M. and Martin, S. M. (1983). Effect of some physical and chemical parameters on the fermentation of cellulose to methane by a coculture system. *Canadian Journal of Microbiology*, **29**, 1475–80.

Lawrence, A. W. and McCarty, P. L. (1965). The role of sulphide in preventing heavy metal toxicity in anaerobic treatment. *Journal of the Water Pollution Control Federation*, **37**, 392–406.

Lawrence, A. W. and McCarty, P. L. (1969). Kinetics of methane fermentation in anaerobic treatment. *Journal of the Water Pollution Control Federation*, **41**, R1–R17.

Leigh, J. A. and Wolfe, R. S. (1983). *Acetogenium kivui* gen. nov., sp. nov., a thermophilic acetogenic bacterium. *International Journal of Systematic Bacteriology*, **33**, 886.

Le Ruyet, P., Dubourguier, H. C. and Albagnac, G. (1984*a*). Homoacetogenic fermentation of cellulose by a coculture of *Clostridium thermocellum* and *Acetogenium kivui*. *Applied and Environmental Microbiology*, **48**, 893–4.

Le Ruyet, P., Dubourguier, H. C. and Albagnac, G. (1984*b*). Thermophilic fermentation of cellulose and xylan by methanogenic enrichment cultures: preliminary characterization of main species. *Systematic and Applied Microbiology*, **5**, 247–53.

Lettinga, G., van Velsen, L., de Zeeuw, W. and Hobma, S. W. (1980). The application of anaerobic digestion to industrial pollution treatment. In *Anaerobic Digestion*, ed. D. A. Stafford, B. I. Wheatley and D. E. Hughes. Applied Science Publishers Ltd, London, pp. 167–86.

Lettinga, G., Hulshoff-Pol, L. W., Wiegant, W., de Zeeuw, W., Hobma, S. W., Grin, P., Roersma, R., Sayed, S. and van Velsen, A. F. W. (1983). Upflow sludge blanket processes. In *Proceedings of the 3rd International Symposium on Anaerobic Digestion*, Boston, USA, pp. 139–58.

Lettinga, G., Hulshoff Pol, L. W., Koster, I. W., Wiegant, W. M., de Zeeuw, W. J., Rinzema, A., Grin, P. C., Roersma, R. E. and Hobma, S. W. (1984). High-rate anaerobic waste-water treatment using the UASB reactor under a whole range of temperature conditions. *Biotechnology and Genetic Engineering Reviews*, **2**, 253–84.

Lettinga, G., Zehnder, A. J. B., Grotenhuis, J. T. C. and Hulshoff Pol, L. W. (1988). *Granular Anaerobic Sludge; Microbiology and Technology*. Pudoc, Wageningen, The Netherlands.

Lotka, A. J. (1925). *Elements of Physical Biology*. Williams & Wilkins, Baltimore.

Lovley, D. R. and Ferry, J. G. (1985). Production and consumption of H_2 during growth of *Methanosarcina* spp. on acetate. *Applied and Environmental Microbiology*, **49**, 247–9.

Lovley, D. R., Greening, R. C. and Ferry, J. G. (1984). Rapidly growing rumen methanogenic organism that synthesizes coenzyme M and has a high affinity for formate. *Applied and Environmental Microbiology*, **48**, 81–7.

Lowe, C. R. (1985). An introduction to the concepts and technology of biosensors. *Biosensors*, **1**, 3–16.

Lupton, F. S. and Zeikus, J. G. (1984). Physiological basis for sulphate-dependent hydrogen competition between sulphidogens and methanogens. *Current Microbiology*, **11**, 7–12.

McBride, B. C. and Wolfe, R. S. (1971). A new coenzyme of methyl-transfer; coenzyme M. *Biochemistry*, **10**, 2317–4.

McInerney, M. J. and Bryant, M. P. (1981). Anaerobic degradation of lactate by syntrophic association of *Methanosarcina barkeri* and *Desulfovibrio* species and effect of H_2 on acetate degradation. *Applied and Environmental Microbiology*, **41**, 346–54.

McInerney, M. J., Bryant, M. P. and Pfennig, N. (1979). Anaerobic bacterium that degrades fatty acids in syntrophic association with methanogens. *Archives of Microbiology*, **122**, 129–35.

McInerney, M. J., Bryant, M. P., Hespell, R. B. and Costerton, J. W. (1981). *Syntrophomonas wolfei* gen. nov., sp. nov., an anaerobic, syntrophic, fatty acid-oxidizing bacterium. *Applied and Environmental Microbiology*, **41**, 1029–39.

Mah, R. A. (1982). Methanogenesis and methanogenic partnerships. *Philosophical Transactions of the Royal Society, London, Series B*, **297**, 599–616.

Mah, R. A. and Smith, M. R. (1981). The methanogenic bacteria. In *The Prokaryotes*, Vol. 1, ed. M. P. Starr, H. Stolp, H. G. Truper, A. Balows and H. G. Schlegel. Springer-Verlag, Berlin and New York, pp. 948–77.

Mah, R. A., Hungate, R. E. and Ohwaki, K. (1977). Acetate, a key intermediate in methanogenesis. In *Microbial Energy Conversion*, ed. H. G. Schlegel and J. Barnea. Pergamon Press, Oxford, pp. 97–106.

Mah, R. A., Smith, M. R. and Baresi, L. (1978). Studies on an acetate-fermenting strain of *Methanosarcina*. *Applied and Environmental Microbiology*, **35**, 1174–84.

Marty, D. G. and Bianchi, A. J. M. (1981). Isolement de deux souches méthanogènes thermophiles appartenant au genre *Methanobacterium*. *Comptes Rendus de l'Académie des Sciences* (Ser. 111), **292**, 41–3.

Martz, R. F., Sebacher, D. I. and White, D. C. (1983). Biomass measurement of methane forming bacteria in environmental samples. *Journal of Microbiological Methods*, **1**, 53–61.

Meynell, P. J. (1982). *Methane: planning a digester*, 2nd edn. Prism Press, Dorchester, Dorset.

Miura, Y., Tanaka, H. and Okazaki, M. (1980). Stability analysis of commensal and mutual relations with competitive assimilation in continuous mixed culture. *Biotechnology and Bioengineering*, **22**, 929–48.

Mosey, F. E. (1982). New developments in the anaerobic treatment of industrial wastes. *Water Pollution Control*, **81**, 540–50.

Mosey, F. E. and Fernandes, X. A. (1984). Mathematical modelling of methanogenesis in sewage sludge digestion. In *Microbiological Methods for Environmental Biotechnology*, ed. J. M. Grainger and J. M. Lynch. Academic Press, London, pp. 159–68.

Mosey, F. E. and Hughes, D. A. (1975). The toxicity of heavy metal ions to anaerobic digestion. *Journal of the Institute of Water Pollution Control*, **1**, 3–24.

Mountfort, D. O. and Asher, R. A. (1979). Effect of inorganic sulphide on the growth and metabolism of *Methanosarcina barkeri* strain DM. *Applied and Environmental Microbiology*, **37**, 670–5.

Mountfort, D. O. and Bryant, M. P. (1982). Isolation and characterisation of an anaerobic syntrophic benzoate-degrading bacterium from sewage sludge. *Archives of Microbiology*, **133**, 249–56.

Mountfort, D. O., Brulla, W. J., Krumholz, L. R. and Bryant, M. P. (1984). *Syntrophus buswellii* gen. nov.: a benzoate catabolizer from methanogenic ecosystems. *International Journal of Systematic Bacteriology*, **34**, 216–7.

Murray, P. A. and Zinder, S. H. (1984). Nitrogen fixation by a methanogenic archaebacterium. *Nature*, **312**, 284–6.

Murray, W. D. and van den Berg, L. (1981). Effects of nickel, cobalt and molybdenum on performance of methanogenic fixed-film reactors. *Applied and Environmental Microbiology*, **42**, 502–5.

Ng, T. K., Ben-Bassat, A. and Zeikus, J. G. (1981). Ethanol production by thermophilic bacteria: fermentation of cellulose substrates by cocultures of *Clostridium thermocellum* and *Clostridium thermohydrosulfuricum*. *Applied and Environmental Microbiology*, **41**, 1337–43.

O'Brien, J. M., Wolkin, R. H., Moench, T. T., Morgan, J. B. and Zeikus, J. G. (1984). Association of hydrogen metabolism with unitrophic or mixotrophic growth of *Methanosarcina barkeri* on carbon monoxide. *Journal of Bacteriology*, **158**, 373–5.

Odelson, D. A. and Breznak, J. A. (1985). Nutrition and growth characteristics of *Trichomitopsis termopsidis*, a cellulolytic protozoan from termites. *Applied and Environmental Microbiology*, **49**, 614–21.
Odum, E. P. (1953). *Fundamentals in Ecology*. Saunders, Philadelphia.
Ohwaki, K. and Hungate, R. E. (1977). Hydrogen utilization by clostridia in sewage sludge. *Applied and Environmental Microbiology*, **33**, 1270–4.
Ollivier, B., Lombardo, A. and Garcia, J. L. (1984). Isolation and characterization of a new thermophilic *Methanosarcina* strain (strain MP). *Annales de l'Institut Pasteur* Microbiologie, **135B**, 187–98.
Patel, G. B. (1984). Characterization and nutritional properties of *Methanothrix concilii* sp. nov., a mesophilic aceticlastic methanogen. *Canadian Journal of Microbiology*, **30**, 1383–96.
Patel, G. B. and Sprott, G. D. (1990). *Methanosaeta concilii* gen. nov., sp. nov. ('*Methanothrix concilii*') and *Methanosaeta thermoacetophila* nom. rev., comb. nov. *International Journal of Systematic Bacteriology*, **40**, 79–82.
Patel, G. B., Khan, A. W. and Roth, L. A. (1978). Optimum levels of sulphate and iron for the cultivation of pure cultures of methanogens in synthetic media. *Journal of Applied Bacteriology*, **45**, 347–56.
Peck, M. W. and Archer, D. B. (1989). Methods for the quantification of methanogenic bacteria. *International Industrial Biotechnology*, **9**, 5–12.
Perski, H.-J., Moll, J. and Thauer, R. K. (1981). Sodium dependence of growth and methane production in *Methanobacterium thermoautotrophicum*. *Archives of Microbiology*, **130**, 319–21.
Pine, M. J. and Barker, H. A. (1954). Studies on the methane bacteria. XI. Fixation of atmospheric nitrogen by *Methanobacterium omelianskii*. *Journal of Bacteriology*, **68**, 589–91.
Pine, M. J. and Barker, H. A. (1956). Studies on the methane fermentation. XII. The pathway of hydrogen in the acetate fermentation. *Journal of Bacteriology*, **71**, 644–8.
Pirt, S. J. (1975). *Principles of Microbe and Cell Cultivation*. Blackwells, Oxford.
Porteous, W. (1983). Sound practice follows from sound theory—the control analysis of Kacser and Burns evaluated. *Trends in Biochemical Science*, **8**, 200–2.
Postgate, J. R. (1984). *The Sulphate-Reducing Bacteria*. Cambridge University Press, Cambridge, UK.
Powell, G. E. (1984). Equalisation of specific growth rates for syntrophic associations in batch culture. *Journal of Chemical Technology and Biotechnology*, **34B**, 97–100.
Powell, G. E. (1985). Stable coexistence of syntrophic associations in continuous culture. *Journal of Chemical Technology and Biotechnology*, **35B**, 46–50.
Powell, G. E. and Archer, D. B. (1989). On-line titration method for monitoring buffer capacity and total volatile fatty acid levels in anaerobic digesters. *Biotechnology and Bioengineering*, **33**, 570–7.
Powell, G. E., Hilton, M. G., Archer, D. B. & Kirsop, B. H. (1983). Kinetics

of the methanogenic fermentation of acetate. *Journal of Chemical Technology and Biotechnology*, **33B**, 209-15.
Powell, G. E., Tatton, M. J. and Archer, D. B. (1985). Dynamics of methanogenesis by mixed cultures. *Proceedings of Advances in Fermentation II*. Turret-Wheatland, Rickmansworth, UK, pp. 110-15.
Prins, R. A., van Nevel, C. J. and Demeyer, D. J. (1972). Pure culture studies of inhibitors for methanogenic bacteria. *Antonie van Leeuwenhoek*, **38**, 281-2.
Rapoport, T. A., Heinrich, R., Jacobasch, G. and Rapoport, S. (1974). A linear steady-state treatment of enzymatic chains. *European Journal of Biochemistry*, **42**, 107-20.
Reddy, C. A., Bryant, M. P. and Wolin, M. J. (1972). Characteristics of S organism isolated from *Methanobacillus omelianskii*. *Journal of Bacteriology*, **109**, 539-45.
Rees, J. F. (1980). The fate of carbon compounds in the landfill disposal of organic matter. *Journal of Chemical Technology and Biotechnology*, **30B**, 161-75.
Richards, S. R. and Archer, D. B. (1989). Anaerobic treatment of high-sulphate molasses wastewaters. In *International Biosystems II*, ed. D. L. Wise. CRC Press, Boca Raton, Florida, pp. 1-28.
Richards, S. R. & Turner, R. (1984). A comparative study of techniques for the examination of biofilms by scanning electron microscopy. *Water Research*, **18**, 767-73.
Robinson, H. D., Barber, C. and Maris, P. J. (1982). Generation and treatment of leachate from domestic wastes in landfills. *Water Pollution Control*, **81**, 465-78.
Robinson, J. A. and Tiedje, J. M. (1982). Kinetics of hydrogen consumption by rumen fluid, anaerobic digester sludge and sediment. *Applied and Environmental Microbiology*, **44**, 1374-84.
Robinson, J. A. and Tiedje, J. M. (1984). Competition between sulphate-reducing bacteria and methanogenic bacteria for H_2 under resting and growing conditions. *Archives of Microbiology*, **137**, 26-32.
Robinson, R. W., Akin, D. E., Nordstedt, R. A., Thomas, M. V. and Aldrich, H. C. (1984). Light and electron microscopic examinations of methane-producing biofilms from anaerobic fixed-bed reactors. *Applied and Environmental Microbiology*, **48**, 127-36.
Rönnow, P. H. and Gunnarsson, L. A. H. (1981). Sulphide-dependent methane production and growth of a thermophilic methanogenic bacterium. *Applied and Environmental Microbiology*, **42**, 580-4.
Roy, F., Albagnac, G. and Samain, E. (1985). Influence of calcium addition on growth of highly purified syntrophic cultures degrading long-chain fatty acids. *Applied and Environmental Microbiology*, **49**, 702-5.
Rozzi, A. and Labellarte, G. (1984). Direct bicarbonate determination in anaerobic digester liquors by measurement of the pressure of carbon dioxide. *Process Biochemistry*, **19**(6), 201-3.
Sanchez Riera, F., Valz-Gianinet, S., Callieri, D. and Sineriz, F. (1982). Use of packed-bed reactor for anaerobic treatment of stillage of sugar cane molasses. *Biotechnology Letters*, **4**, 127-32.

Scherer, P. and Sahm, H. (1981a). Effect of trace elements and vitamins on the growth of *Methanosarcina barkeri*. *Acta Biotechnologica*, **1**, 57–65.

Scherer, P. and Sahm, H. (1981b). Influence of sulphur-containing compounds on the growth of *Methanosarcina barkeri* in a defined medium. *European Journal of Applied Microbiology and Biotechnology*, **12**, 28–35.

Scherer, P., Lippert, H. and Wolff, G. (1983). Composition of the major elements and trace elements of 10 methanogenic bacteria determined by inductively coupled plasma emission spectrometry. *Biological Trace Element Research*, **5**, 149–63.

Schink, B. (1984a). *Clostridium magnum* sp. nov., a non-autotrophic homoacetogenic bacterium. *Archives of Microbiology*, **137**, 250–5.

Schink, B. (1984b). Fermentation of 2,3-butanediol by *Pelobacter carbinolicus* sp. nov. and *Pelobacter propionicus* sp. nov., and evidence for propionate formation from C2 compounds. *Archives of Microbiology*, **137**, 33–41.

Schink, B. (1985a). Mechanism and kinetics of succinate and propionate degradation in anoxic freshwater sediments and sewage sludge. *Journal of General Microbiology*, **131**, 643–50.

Schink, B. (1985b). Inhibition of methanogenesis by ethylene and other unsaturated hydrocarbons. *FEMS Microbiology Ecology*, **31**, 63–8.

Schink, B. (1985c). Degradation of unsaturated hydrocarbons by methanogenic enrichment cultures. *FEMS Microbiology Ecology*, **31**, 69–77.

Schink, B. and Pfennig, N. (1982). Fermentation of trihydroxybenzenes by *Pelobacter acidigallici* gen. nov. sp. nov., a new strictly anaerobic, non-sporeforming bacterium. *Archives of Microbiology*, **133**, 195–201.

Schönheit, P. and Thauer, R. K. (1980). L-alanine, a product of cell wall synthesis in *Methanobacterium thermoautotrophicum*. *FEMS Microbiology Letters*, **9**, 77–80.

Schönheit, P., Moll, J. and Thauer, R. K. (1980). Growth parameters (K_s, μ_{max}, Y_s) of *Methanobacterium thermoautotrophicum*. *Archives of Microbiology*, **127**, 59–65.

Schönheit, P., Kristjansson, J. K. and Thauer, R. K. (1982). Kinetic mechanism for the ability of sulphate reducers to out-compete methanogens for acetate. *Archives of Microbiology*, **132**, 285–8.

Scott, R. I., Williams, T. N., Whitmore, T. N. and Lloyd, D. (1983). Direct measurement of methanogenesis in anaerobic digestors by membrane inlet mass spectrometry. *European Journal of Applied Microbiology and Biotechnology*, **18**, 236–41.

Seely, R. J. and Fahrney, D. E. (1984). Levels of cyclic-2,3-diphosphoglycerate in *Methanobacterium thermoautotrophicum* during phosphate limitation. *Journal of Bacteriology*, **160**, 50–4.

Senior, E. and Balba, M. T. M. (1983). The microbiology and biochemistry of landfill biotechnology. Conference: 'Reclamation 83', Grays, Essex, 26–29 April 1983. *Industrial Seminars Ltd*, pp. 337–52.

Senior, E. and Balba, M. T. M. (1984). The use of single-stage and multi-stage fermenters to study the metabolism of xenobiotic acid and naturally occurring molecules by interacting microbial associations. In *Microbiological Methods for Environmental Biotechnology*, ed. J. M. Grainger and J. M. Lynch. Academic Press, London, pp. 275–93.

Sharak-Genthner, B. R., Davis, C. L. and Bryant, M. P. (1981). Features of rumen and sewage sludge strains of *Eubacterium limosum*, a methanol- and H_2 . CO_2-utilizing species. *Applied and Environmental Microbiology*, **42**, 12–19.

Shelton, D. R. and Tiedje, J. M. (1984). Isolation and partial characterisation of bacteria in an anaerobic consortium that mineralizes 3-chlorobenzoic acid. *Applied and Environmental Microbiology*, **48**, 840–8.

Sibold, L., Pariot, D., Bhatnagar, L., Henriquet, M. and Aubert, J.-P. (1985). Hybridization of DNA from methanogenic bacteria with nitrogenase structural genes (nifHDK). *Molecular and General Genetics*, **200**, 40–6.

Sleat, R. and Robinson, J. P. (1984). The bacteriology of anaerobic degradation of aromatic compounds. *Journal of Applied Bacteriology*, **57**, 381–94.

Sleat, R., Mah, R. A. and Robinson, R. (1985). *Acetoanaerobicum noterae* gen. nov., sp. nov.: an anaerobic bacterium that forms acetate from H_2 and CO_2. *International Journal of Systematic Bacteriology*, **35**, 10–15.

Smith, M. R. and Mah, R. A. (1978). Growth and methanogenesis by *Methanosarcina* strain 227 on acetate and methanol. *Applied and Environmental Microbiology*, **36**, 870–9.

Smith, M. R., Zinder, S. H. and Mah, R. A. (1980). Microbial methanogenesis from acetate. *Process Biochemistry*, **15**(4), 34–9.

Smith, P. H. (1966). The microbial ecology of sludge methanogenesis. *Developments in Industrial Microbiology*, **7**, 156–60.

Smith, P. H. and Hungate, R. E. (1958). Isolation and characterisation of *Methanobacterium ruminantium* n.sp. *Journal of Bacteriology*, **75**, 713–18.

Sowers, K. R., Baron, S. F. and Ferry, J. G. (1984). *Methanosarcina acetivorans* sp. nov., an acetotrophic methane-producing bacterium isolated from marine sediments. *Applied and Environmental Microbiology*, **47**, 971–8.

Speece, R. E., Parkin, G. F. and Gallagher, D. (1983). Nickel stimulation of anaerobic digestion. *Water Research*, **17**, 677–83.

Sprott, G. D. and Jarrell, K. F. (1981). K^+, Na^+ and Mg^{2+} content and permeability of *Methanospirillum hungatei* and *Methanobacterium thermoautotrophicum*. *Canadian Journal of Microbiology*, **27**, 444–51.

Sprott, G. D., Jarrell, K. F., Shaw, K. M. and Knowles, R. (1982). Acetylene as an inhibitor of methanogenic bacteria. *Journal of General Microbiology*, **128**, 2453–62.

Stadtman, T. C. and Barker, H. A. (1949). Studies on the methane fermentation VII. Tracer experiments on the mechanism of methane formation. *Archives of Biochemistry*, **21**, 256–64.

Standard Methods for the Examination of Water and Wastewater (1960). American Public Health Association, New York.

Strayer, R. F. and Tiedje, J. M. (1978). Application of the fluorescent-antibody technique to the study of a methanogenic bacterium in lake sediment. *Applied and Environmental Microbiology*, **35**, 192–8.

Stumm, C. K., Gijzen, H. J. and Vogels, G. D. (1982). Association of methanogenic bacteria with bovine rumen ciliates. *British Journal of Nutrition,* **47,** 95–9.

Suflita, J. M., Horowitz, A., Shelton, D. R. and Tiedje, J. M. (1982). Dehalogenation: a novel pathway for the anaerobic biodegradation of haloaromatic compounds. *Science,* **218,** 1115–17.

Switzenbaum, M. S. (1983). Anaerobic fixed film wastewater treatment. *Enzyme and Microbial Technology,* **5,** 242–50.

Szendrey, L. M. (1983). Startup and operation of the Bacardi Corporation anaerobic filter. In *Proceedings of the 3rd International Symposium on Anaerobic Digestion,* Boston, USA, pp. 365–77.

Szewzyck, U., Szewzyk, R. and Schink, B. (1985). Methanogenic degradation of hydroquinone and catechol via reductive dehydroxylation to phenol. *FEMS Microbiology Ecology,* **31,** 79–87.

Tatton, M. J., Archer, D. B., Powell, G. E. and Parker, M. L. (1989). Methanogenesis from ethanol by defined mixed continuous cultures. *Applied and Environmental Microbiology,* **55,** 440–5.

Taylor, G. T. (1982). The methanogenic bacteria. In *Progress in Industrial Microbiology,* Vol. 16, ed. M. J. Bull. Elsevier, Amsterdam, pp. 231–329.

Thiele, J. H. and Zeikus, J. G. (1988). Control of interspecies electron flow during anaerobic digestion: significance of formate transfer versus hydrogen transfer during syntrophic methanogenesis in flocs. *Applied and Environmental Microbiology,* **54,** 20–9.

Toerien, D., Siebert, M. L. and Hattingh, W. H. J. (1967). The bacterial nature of the acid-forming phase of anaerobic digestion. *Water Research,* **1,** 497–507.

Toerien, D., Thiel, P. G. and Hattingh, M. M. (1968). Enumeration, isolation and identification of sulphate-reducing bacteria of anaerobic digestion. *Water Research,* **2,** 505–13.

Traore, A. S., Gaudin, C., Hatchikian, C. E., Le Gall, J. and Belaich, J.-P. (1983a). Energetics of growth of a defined mixed culture of *Desulfovibrio vulgaris* and *Methanosarcina barkeri*: maintenance energy coefficient of the sulphate-reducing organism in the absence and presence of its partner. *Journal of Bacteriology,* **155,** 1260–4.

Traore, A. S., Fardeau, M.-L., Hatchikian, C. E., Le Gall, J. and Belaich, J.-P. (1983b). Energetics of growth of a defined mixed culture of *Desulfovibrio vulgaris* and *Methanosarcina barkeri*: interspecies hydrogen transfer in batch and continuous cultures. *Applied and Environmental Microbiology,* **46,** 1152–6.

Valcke, D. and Verstraete, W. (1983). A practical method to estimate the acetoclastic methanogenic biomass in anaerobic sludges. *Journal of the Water Pollution Control Federation,* **55,** 1191–5.

van Beelen, P., Dijkstra, A. C. and Vogels, G. D. (1983a). Quantitation of coenzyme F420 in methanogenic sludge by the use of reversed-phase high-performance liquid chromatography and a fluorescence detector. *European Journal of Applied Microbiology and Biotechnology,* **18,** 67–9.

van Beelen, P., Geerts, W. J., Pol, A. and Vogels, G. D. (1983b). Quantification of coenzymes and related compounds from methanogenic

bacteria by high-performance liquid chromatography. *Analytical Biochemistry*, **131**, 285–90.
van Bruggen, J. J. A., Zwart, K. B., van Assema, R. M., Stumm, C. K. and Vogels, G. D. (1984). *Methanobacterium formicicum*, an endosymbiont of the anaerobic ciliate *Metopus striatus* McMurrich. *Archives of Microbiology*, **139**, 1–7.
van den Berg, L. and Lentz, C. P. (1981). Performance and stability of the anaerobic contact process as affected by waste composition, inoculation and solids retention time. *Proceedings of the 35th Industrial Waste Conference*, Purdue University, Lafayette, Indiana. Ann Arbor Science, Michigan, USA, pp. 496–501.
van den Berg, L., Patel, G. B., Clark, D. S. and Lentz, C. P. (1976). Factors affecting rate of methane formation from acetic acid by enriched methanogenic cultures. *Canadian Journal of Microbiology*, **22**, 1312–19.
van den Berg, L., Lamb, K. A., Murrey, W. D. and Armstrong, D. W. (1980). Effects of sulphate, iron and hydrogen on the microbiological conversion of acetic acid to methane. *Journal of Applied Bacteriology*, **48**, 437–47.
Veal, D. A. and Lynch, J. M. (1984). Biochemistry of cellulose breakdown by mixed cultures. *Biochemistry Society Transactions*, **12**, 1142–4.
Vogel, T. M. and McCarty, P. L. (1985). Biotransformation of tetrachloroethylene to trichloroethylene, dichloroethylene, vinyl chloride, and carbon dioxide under methanogenic conditions. *Applied and Environmental Microbiology*, **49**, 1080–3.
Vogels, G. D., Hoppe, W. F. and Stumm, C. K. (1980). Association of methanogenic bacteria with rumen ciliates. *Applied and Environmental Microbiology*, **40**, 608–12.
Vogels, G. D., van der Drift, C., Stumm, C. K., Keltjens, J. T. M. and Zwart, K. B. (1984). Methanogenesis: surprising molecules, microorganisms and ecosystems. *Antonie van Leeuwenhoek*, **50**, 557–67.
Volterra, V. (1927). Variazioni e fluttuazioni del numero di individui in specie animali conviventi. *Atti Reale Accademia nazionale dei Lincei, Serie VI, Memorie della Classe di Scienze Fisiche, Matematiche e Naturali*, **11**, 31–112; translated in *Applicable Mathematics of Non-Physical Phenomena* (1982), ed. F. Oliveira-Pinto and B. W. Conolly. Ellis Horwood, Chichester, pp. 23–115.
Wang, Y.-T., Suidan, M. T. and Pfeffer, J. T. (1984). Anaerobic biodegradation of indole to methane. *Applied and Environmental Microbiology*, **48**, 1058–60.
Ward, D. M., Mah, R. A. and Kaplan, I. R. (1978). Methanogenesis from acetate: a non-methanogenic bacterium from an anaerobic acetate enrichment. *Applied and Environmental Microbiology*, **35**, 1185–92.
Weber, H., Kulbe, K. D., Chmiel, H. and Trösch, W. (1984). Microbial acetate conversion to methane: kinetics, yields and pathways in a two-step digestion process. *Applied Microbiology and Biotechnology*, **19**, 224–8.
Wellinger, A. and Wuhrmann, K. (1977). Influence of sulphide compounds on the metabolism of *Methanobacterium* strain AZ. *Archives of Microbiology*, **115**, 13–17.

Wheatley, A. D., Winstanley, C. I. and Johnson, K. A. (1987). The development of control procedures for the anaerobic treatment of strong and variable industrial wastes. In *Anaerobic Digestion*: Results of Research and Demonstration projects, ed. M. T. Ferranti, G. L. Ferrero and P. L'Hermite. Elsevier Science Publishers, London, pp. 250–60.
Whitmore, T. N. and Lloyd, D. (1986). Mass spectrometric control of the thermophilic anaerobic digestion process based on levels of dissolved hydrogen. *Biotechnology Letters*, **8**, 203–8.
Whitmore, T. N., Lazzari, M. and Lloyd, D. (1985). Comparative studies of methanogenesis in thermophilic and mesophilic anaerobic digesters using membrane inlet mass spectrometry. *Biotechnology Letters*, **7**, 283–8.
Wiegel, J., Braun, M. and Gottschalk, G. (1981). *Clostridium thermoautotrophicum* species novum, a thermophile producing acetate from molecular hydrogen and carbon dioxide. *Current Microbiology*, **5**, 255–60.
Wieringa, K. T. (1940). The formation of acetic acid from carbon dioxide and hydrogen by anaerobic spore-forming bacteria. *Antonie van Leeuwenhoek*, **6**, 251–62.
Williamson, M. (1972). *The Analysis of Biological Populations*. Edward Arnold, London.
Winter, J. U. and Wolfe, R. S. (1980). Methane formation from fructose by syntrophic associations of *Acetobacterium woodii* and different strains of methanogens. *Archives of Microbiology*, **124**, 73–9.
Wolin, M. J. (1976). Interactions between H_2-producing and methane-producing species. In *Microbial Production and Utilization of Gases (H_2, CH_4, CO)*, ed. G. Schlegel, G. Gottschalk and N. Pfenning. E. Goltze KG, Göttingen, pp. 141–50.
Wolin, M. J. (1982). Hydrogen transfer in microbial communities. In *Microbial Interactions and Communities*, ed. A. T. Bull and J. H. Slater. Vol. 1, Academic Press, London, pp. 323–56.
Zabranska, J., Schneiderova, K. and Dohanyos, M. (1985). Relation of coenzyme F420 to the activity of methanogenic microorganisms. *Biotechnology Letters*, **7**, 547–52.
Zehnder, A. J. B. and Koch, M. E. (1983). Thermodynamic and kinetic interactions of the final steps in anaerobic digestion. In *Anaerobic Waste Water Treatment*, ed. W. J. van den Brink. TNO Corporate Communication Department, The Hague, pp. 86–96.
Zehnder, A. J. B. and Wuhrmann, K. (1977). Physiology of a *Methanobacterium* strain AZ. *Archives of Microbiology*, **111**, 199–205.
Zehnder, A. J. B., Huser, B. A., Brock, T. D. and Wuhrmann, K. (1980). Characterization of an acetate-decarboxylating, non-hydrogen-oxidizing methane bacterium *Archives of Microbiology*, **124**, 1–11.
Zeikus, J. G. (1980). Microbial Populations in Digesters. In *Anaerobic Digestion*, ed. D. A. Stafford, B. I. Wheatley and D. E. Hughes. Applied Science Publishers Ltd, London, pp. 61–89.
Zeikus, J. G., Lynd, L. H., Thompson, T. E., Krzycki, J. A., Weimer, P. J. and Hegge, P. W. (1980). Isolation and characterization of a new, methylotrophic, acidogenic anaerobe, the Marburg strain. *Current Microbiology*, **3**, 381–6.

Zeikus, J. G., Wellstein, A. L. and Kirk, T. K. (1982). Molecular basis for the biodegradative recalcitrance of lignin in anaerobic environments. *FEMS Microbiology Letters*, **15**, 193–7.

Zinder, S. H. (1984). Microbiology of anaerobic conversion of organic wastes to methane: recent developments. *American Society of Microbiology News*, **50**, 294–8.

Zinder, S. H. and Koch, M. (1984). Non-aceticlastic methanogenesis from acetate: acetate oxidation by a thermophilic syntrophic coculture. *Archives of Microbiology*, **138**, 263–72.

Zinder, S. H. and Mah, R. A. (1979). Isolation and characterization of a thermophilic strain of *Methanosarcina* unable to use H_2–CO_2 for methanogenesis. *Applied and Environmental Microbiology*, **38**, 996–1008.

Zinder, S. H., Cardwell, S. C., Anguish, T., Lee, M. and Koch, M. (1984). Methanogenesis in a thermophilic (58°C) anaerobic digester: *Methanothrix* sp. as an important aceticlastic methanogen. *Applied and Environmental Microbiology*, **47**, 796–807.

3 The treatment of agricultural wastes

P. N. Hobson

University of Aberdeen, UK

3.1	Introduction	94
3.2	The role of anaerobic digestion	95
3.3	The characteristics of farm wastes	97
3.3.1	Volumes produced	97
3.3.2	The physical characteristics of farm wastes	98
3.3.3	The chemical composition of farm wastes	99
3.4	Process design	102
3.5	Types of digester	104
3.5.1	Batch digesters	105
3.5.2	The stirred-tank, continuous-flow, digester	108
3.5.3	The tubular or plug-flow digester	110
3.5.4	Retained-biomass digesters	111
	(a) The anaerobic filter	113
	(b) The flocculated-biomass digesters	114
	(c) The contact digester	114
	(d) Two-stage, two-phase and hybrid digesters	115
3.6	Digester operation	117
3.6.1	Retention time	117
3.6.2	Solids concentration	119
3.6.3	Temperature	120
3.6.4	Inhibitory materials	122
3.6.5	Start-up	123
3.7	Digester construction	124
3.8	Ancillary equipment	125
3.8.1	Gas storage	125
3.8.2	Heating the digester	127
3.8.3	Mixing	128
3.8.4	Feeding	128

3.9	**Residues after digestion**	128
3.10	**Modelling**	130
3.11	**Conclusions**	131
3.12	**References**	132
3.13	**Bibliography**	137

3.1 Introduction

Agricultural wastes are the largest single source of waste and potential pollution in Europe. The main problem is the excreta produced by intensively-reared farm animals. Economies of scale have led to a change in farming practice in many countries from pasture feeding to 'intensive', indoor, rearing of animals and birds. It is this change that has required the management of farm wastes. If cattle are kept for all, or most, of the year grazing in fields or on open ranch land then excreta deposited on the fields are easily recycled as fertilizer to the grass by natural processes without offence. In Britain and other countries where animals are housed during the winter on mixed farms the animal excreta can, after storage if necessary, be returned as fertilizer to the grass or arable land of the farm. The only treatment is natural decay of excreta and straw in the midden, and the resultant 'farm yard manure' (FYM) can be spread on the land with little or no offence to the public or pollution of water supplies. Similarly, excreta from a few pigs or hens kept in houses or in fields provide few problems in disposal.

In a typical intensive unit, cattle, pigs, poultry or ducks are kept in housing all the year round, and in very large numbers. There are frequently hundreds or thousands of cattle or pigs and millions of birds raised in houses on a very small area of ground.

The number of this type of installation has grown rapidly in recent years. In these cases there is too much waste to be safely disposed of on the farm's land. The disposal of wastes as fertilizer to adjacent farms without intensive units, even if possible, is also rarely satisfactory. Frequent problems are complaints about the smell and pollution resulting from run-off to water courses. These difficulties are due to both the quantities and the nature of the wastes. Small numbers of animals are kept on straw which absorbs much of the liquid waste and a certain amount of the odour. Intensive units use slatted-floor or similar housing where the mixed faeces and urine and the spilled drinking and the washing waters form a slurry in under-floor tanks. This slurry ferments during storage in these tanks and when disturbed for land spreading it can give off explosive, malodorous and poisonous gases. Spreading, particularly by the rain-gun type of spreader or the pressurized tanker, is very effective at creating fine liquid droplets and noxious odours which are carried long distances by the wind. Serious pollution by run-off is also much more likely because the waste is more dilute than FYM. Run-off will occur when the land is rain-soaked, has a high water table, or is frozen. Land spreading is not possible at all times of the year. Thus, there is need for new methods of treating the waste from intensive farming units, and one method is anaerobic digestion. During the last few years much experience has been gained in designing and running anaerobic digesters. Equipment is more reliable than were the experimental plants of some years ago. Low oil prices have diminished the value of the biogas produced as a farm fuel but anaerobic digestion as a pollution-control process is now more important. Public awareness of pollution has increased the number of regulations to prevent environmental damage.

In a short chapter such as this it is impossible to discuss all the work that has been done on the anaerobic digestion of farm waste. A short list of references to papers and books is therefore given as a guide to further reading and to further information. Some important or representative references are given in the text, but, overall, the discussion has been generalized.

3.2 The role of anaerobic digestion

Traditionally, wastewater treatment has used aerobic biological processes. The pollution potential of dilute wastewaters such as settled sewage can easily be reduced by aeration. Animal excreta, even from intensive units, are, however, thick slurries and complete pollution control by aeration of farm slurries is economically impossible. The energy requirements and complexity make such plants impractical. Aeration can be satisfactory as a means of eliminating odour from stored slurry before spreading but it is expensive. Aeration and aerobic treatment may also be used as a secondary treatment to polish effluents after other types of treatment. Thus anaerobic digestion is the preferred pollution-control system. There are additional advantages:

(1) virtual elimination of pathogens (both plant and animal) and of weed seeds;
(2) an improvement in the fertilizer value of the waste;
(3) the production of a useful byproduct, methane (biogas).

The fertilizer and methane are particularly important in countries with agriculture-based economies and shortages of fuels. The value of the methane is often over-estimated in industrial countries with power networks.

Anaerobic digestion by itself is not a complete treatment and the treated effluents are not usually suitable for discharge to water courses. This is partly because of the nature of the process and partly because the starting material is so high in pollutants (frequently several hundred times more polluting than sewage) that a process efficiency of well over 99% has to be attained before the pollutants are reduced to concentrations suitable for river discharge. However, much of the polluting material can be removed (50–90% depending on what is measured) economically and the remainder is stabilized so that the problems of odour and dangerous gas production on storage and spreading are virtually eliminated. The problems from decomposition and run-off of the spread slurry are also much reduced. By combining digester systems with other technology such as solids separation and secondary treatments it is possible to produce effluents suitable for river discharge or recycling within the farm. However, this significantly adds to the costs and it is rarely worthwhile unless local conditions and legislation are very stringent.

Anaerobic digestion is also becoming more popular in the food-processing industries. Pollution by industry is more strictly controlled than is pollution by farms and reduction in the strength of processing-factory wastes is usually essential. Charges for local disposal are based on the strength of the wastes so it is frequently economic to use anaerobic pretreatment at the factory before discharging the waste to sewers (see Chapter 5).

Much of the interest in anaerobic digestion was a few years ago based on the possibility of generating power which could be used elsewhere and thus give a return on the costs of pollution control. The impetus in industrial countries came from the 1970–74 oil crisis when the cost of oil increased sharply. Much research was conducted to develop the process as an energy-supply system. Fossil fuel prices are now half those of 1980 and the emphasis has shifted toward anaerobic digestion as a cost-effective pollution-control technology. However, energy generation can still be an important consideration in rural economies. The biogas can be used to produce high-grade energy in the form of heat, electricity or motive power, singly or in combination. This energy can compensate for a lack of other easily-accessible sources of energy such as wood, or replace these rapidly-diminishing supplies.

3.3 The characteristics of farm wastes

3.3.1 Volumes produced

Normally, the initial design for a digester treating farm-animal wastes is based on the figure for waste production from one animal and the gas production from a unit amount of that waste. Some figures are given in Table 3.1. These figures give only an approximate guide as they are averages from the work of the author and others (see, for instance, Palz and Cartier, 1980; Hobson et al., 1981; Loehr, 1984; Hobson and Robertson, 1977, for some lists and references). Actual gas production is affected by details of animal feeding, type of feed, age, category (e.g. beef or dairy cow) and digestion conditions. In the case of poultry litter no averages were found in the literature and figures for a particular broiler litter are given. Differences in feed composition and in feeding regime (e.g. feeding *ad lib.* or controlled amount) can account for some of the differences in quantities of excreta and gas production to be found. Major differences between dairy and beef cattle and between cattle in different countries are reported. Much careful consideration of the data available is therefore necessary before the design of a digester is finalized. While the likely volume of excreta produced by the animals on a farm can be estimated from average data such as those in Table 3.1 or by using figures for conversion of food to faeces by different types of animals (e.g. Hobson and Robertson, 1977), the actual volume must be checked by on-site measurements. For instance, the computed volume can be doubled by inflow of water as described below. There have been various attempts at estimating the amounts of wastes produced nationally. These figures have been summarized in books (see Bibliography) and have been used to give some idea of the magnitude of the farm-waste problem

Table 3.1 Excreta production from farm animals and biogas production from the excreta in mesophilic digestions

Animal	Excreta/animal day (kg)	DM in excreta (%)	Biogas (litre/kg VS)	Methane in biogas (%)
Calf	7	13	—	—
Beef heifer	21	13	220–300	55–60
Beef cow	28	9–14	—	—
Dairy cow	45	13	220–400	55–60
Growing-finishing pig	5	6–10	300–400	68–70
Sow and litter	15	4	—	—
Laying hen	0·1	30–50	350–450	65–70
Broiler	0·06	—	—	—
Broiler litter	a	50–80	250–370	60

DM, Dry Matter (Total Solids, TS), approximate values in waste as excreted.
VS, Volatile Solids (organic matter). This varies, but 70% VS of the TS is about average for wastes other than litters.
a Broiler litter, excreta plus sawdust in this case. The excreta are as above but the amount of waste generated depends on the amount of litter per bird and the time that the birds are on the litter.

and the amount of energy and fertilizer that could be obtained from the wastes. For instance, 120, 219, 740 and 1951 million tonnes were quoted as the annual production of wet animal excreta in the UK, France, the other nine EC countries of the time, and the USA. Excreta from pigs in the UK and USA were 11 and 22 million tonnes and for poultry in the UK were 4·5 million tonnes per year. Cereal residues in England and Wales were 10 million and in the USA 546 million tonnes per year. (Figures from Palz and Cartier, 1980; Hobson et al., 1981.) Such figures have to be used with care in estimating energy and other scenarios; the wastes are produced on farms of different sizes and some may not be available for digestion or may not be causing pollution problems.

3.3.2 The physical characteristics of farm wastes

The data in Table 3.1 show that farm-animal wastes are produced as either thick slurries or wet solids. The feed of the animal is one of the factors that affects the solids content of the excreta. With pigs, for instance, a liquid (e.g. whey) diet will produce excreta of 5–6% solids while a dry meal diet will give excreta of 9–10% solids. Addition of some of the other types of wastes likely to be found on farms, crop wastes or bedding materials, will increase the solids content of excreta, as will the natural drying of excreta in a warm poultry house (the reason for the 50–80% figures in Table 3.1). Some other types of farm wastes may tend to dilute the excreta. For example, in animal houses and milking parlours animals and equipment are washed down, and the wastewater often runs into the excreta tanks, as do splashings and drips from drinkers. Rainwater also commonly gets into slurries. Dilution of excreta is a common phenomenon. Further reductions in solids content of particularly pig slurry are caused by settlement of the solids, as the slurry moves slowly along the under-floor channels or during storage. Once solids have settled out of slurries they become very difficult to move and solids accumulate in the systems. The solids in the slurry feeding a digester can, then, be lower than the calculated value and the main source of biogas in slurries is the solids.

3.3.3 The chemical composition of farm wastes

Much of the slurries from animals is excretal and the slurry formed is basically similar in chemical composition to the primary sludges of a municipal sewage plant. Not all the compounds, of similar chemical analysis in different sludges, are equally available for bacterial metabolism. Some analyses of sludges and slurries are given in Table 3.2. Most animal wastes contain more fibre than sewage sludge. Cellulose in sewage sludge is mainly residues of prepared and cooked vegetables used as human food, along with tissue papers and other materials made of delignified, manufactured, plant fibres. Such lignin-free materials are relatively easily degraded by microbes. The cellulose and hemicellulose in animal excreta, while of similar general analysis to sewage sludge, contain highly-lignified residues of the leaves, stalks and grain husks of animal feeds, which have resisted microbial degradation in the animal rumen or hind-gut. Since lignin is non-degradable anaerobically and its presence prevents degradation of plant carbohydrates these fibres are much more resistant to further microbial degradation than are the fibres in the sewage sludge. Residual carbohydrates, and the volatile fatty acids formed by

Table 3.2 The compositions of solids from some farm and sewage wastes (% dry weight)

Waste from	N	Fat	ADF	NDF	Lignin	Ash
Fattening cattle	1·6	2·9	44·5	60·3	11·8	8·2
Dairy cattle	2·3	6·5	41·0	56·7	13·8	14·1
Pigs	7·4	13·7	23·6	45·2	8·1	14·0
Poultry	5·1	6·8	20·1	37·2	6·5	27·5

	Cellulose	Hemicellulose	Lignin	Fat	Protein	Ash
Straw	42·5	26·8	10·3	—	4	12·5
Sewage sludge	3·8	3·2	5·8	34·4	27·1	24·1

N—organic nitrogen (in compounds such as proteins).
ADF—Acid Detergent Fibre (cellulose and lignin).
NDF—Neutral Detergent Fibre (cellulose, hemicellulose and lignin).
Animal excreta figures are from the author's work and apply to particular samples. Variations can be caused by animal feed, waste collection methods, etc. 'As excreted' compositions may also differ from these figures which were for wastes collected from animal-houses. The liquid fractions of the wastes could contain from about 1000–9000 mg VFA/litre and from 500–5000 mg NH_3–N/litre depending on type of waste and time since excretion.
Cereal (barley) straw figures are from Morrison (1979) and others; sewage sludge is from Pohland (1962; quoted by Hobson et al., 1981). These apply to particular samples except for the straw protein which is a generalized figure.

fermentation of carbohydrates in the animal gut or during storage of the excreta, are the principal sources of biogas in animal excreta feedstocks. There is little fat in such wastes (Table 3.2). Carbohydrates and fats are the usual principal primary sources of carbon for methane production (see Chapter 2). While the overall compositions of all livestock excreta are similar, the available carbohydrate varies with the type of animal and its digestive system.

The microbes of the rumen remove most of the more easily available feed carbohydrates. The residues in ruminant faeces are only slowly and incompletely metabolized by the digester bacteria. Microbial digestion of feed in the pig gut is less extensive and more biodegradable material is left for the digester bacteria. The effects of feed structure were shown, for instance, by Stanogias et al. (1985) who found very large differences in specific methane production from faeces of pigs of the same age but fed on rations containing different types of vegetable fibre.

Thus, the degradation of cellulose and hemicellulose in vegetable fibres is controlled by the physical structure of the fibres and the extent of their lignification. Bacteria utilize organic carbon compounds as substrates, but for growth to occur there must also be sufficient nitrogen, sulphur and phosphorus as well as some trace elements as salts. Suitable bacterial nutrients, although not always in the optimum proportions, are present in animal excreta and these materials will digest without further additions. The nitrogenous and other additional nutrients in farm sludges are derived from intestinal bacteria, epithelial cells and secretions. Ammonia is derived from urine. Materials such as straw may be deficient in nitrogen, so for optimum digestion they need to be mixed with a material with excess nitrogen such as animal excreta. This is fortunate, as often the materials are produced in close proximity or are already mixed together as when straw is used for animal bedding, or straw, wood shavings, corn cobs, etc., are used as the basis for poultry litters. It is usual to mix all types of wastes together in anaerobic digesters in countries with rural economies. Optimum proportions of carbon to nitrogen have been suggested for digester feedstocks, but there are problems due to the availability of nutrients to the bacteria. This availability is not easily determined by analysis. Hills (1979) fed mesophilic digesters with various mixtures of screened dairy manure ($C:N = 8:1$) plus a prepared cellulose powder. The cellulose in this mixture would be almost entirely degradable and maximum methane production was found at a $C:N$ ratio of $25:1$, where C was total C minus lignin C: i.e. carbon apparently potentially available to the bacteria. The results were complicated, however, and maximum methane production did not correspond with maximum biogas ($CO_2 + CH_4$) production, as others have found. The $C:N$ ratio of animal wastes can be changed by addition of vegetable wastes and this can improve the ratio for microbial growth and fermentation. However, the main object of such additions is to get rid of a polluting (or potentially polluting) waste, with some economic return in biogas, so supply patterns mean that it might not be possible to produce mixtures of optimum $C:N$ ratio. Robbins et al. (1983) found that mixtures of 5% Total Solids (TS) dairy manure, plus prepared cellulose powder (which represented 40% of the Volatile Solids) gave the maximum methane yield. Barley straw added to a pig slurry to give a 4·3% TS slurry (from a 3·3% TS) was found to be only about 35% digestible even when the retention time was increased up to 20 days.

This amount of straw increased the gas production without appreciably changing the methane content of the gas (69%). The theoretical gas produced from carbohydrates has 50% of methane and carbon dioxide and so large amounts of added carbohydrate could decrease the methane content of the gas produced from the usual faecal mixture of carbohydrates, fats and proteins (Hobson, 1979). Badger et al. (1979) conducted experiments on the digestion of crop residues alone and obtained yields from about 300–500 litres of gas (69–70% methane) per kg of grass, cereals or straw. The amount of gas was proportional to the structure of the carbohydrate. Biogas has also been produced from various tropical crop wastes and weeds, usually added to the normal cattle-waste feedstock of rural digesters (e.g. Dar and Tandon, 1987; Sharma et al., 1989) and from seaweeds and microalgae (e.g. Yang, 1981; Brouard et al., 1983). Waste cardboard, sawdust and newspaper have been digested with pig waste (Wong and Cheung, 1989). Waste potatoes were added to pig waste in other experiments to provide extra carbohydrate (and to test the possibility of disposal of undersized and rotten potatoes). The principal carbohydrate in potatoes is starch, which is readily digested at 10 day's retention time (Summers and Bousfield, 1980). Silage liquid, a highly-polluting mixture of dissolved fermentation acids and constituents of plant juices which seeps from silage heaps and silos, can also be readily digested with pig waste or by itself in a filter digester (Summers and Bousfield, 1980; Barry and Colleran, 1982).

3.4 Process design

There are three factors to consider:

(1) the amount of waste;
(2) the nature of the waste;
(3) the location of the digester.

The stirred-tank digester is the usual type for farm use as it is suitable for slurry and high solids feedstocks. The main substrate is the slurry solids and solids content is used to determine gas production. If solids are low gas production will be low per unit volume of waste treated. As noted in the previous section, the outline size of the digester can be calculated from the number of animals. The amount of dilution of the excreta from rain and wash water and other sources has then to be estimated. This will give an idea of the possible composition of the digester feedstock. If it appears that the slurry will be too dilute for an economic digester in terms of size and production of sufficient gas to at least heat the digester, then a change in farm buildings or farm practice will have to be considered. Extraneous water will have to be diverted to drains if it is rainwater, or to some form of treatment plant if it contains pollutants. Dilute wastes, for example dairy washings which contain milk and some animal excreta from the milking parlour, would be better treated separately from the excreta by lagoon, aerobic ditch, trickling filter, or a retained-biomass anaerobic system. The process used should be as simple as possible. If the animal excreta have been very diluted and the diluent cannot be removed then digesters of types other than the stirred-tank, with or without initial separation of solids, should be considered.

Although on paper a farm may appear to produce a slurry of suitable solids content and volume for running a stirred-tank digester it may not be possible to get this slurry to the digester in a uniform flow of relatively constant solids content. Farm slurry systems are usually built to be emptied at intervals by, for instance, release of a dam which in theory flushes solids from the channels, or after mixing of tank contents by tractor tankers or mobile mixers. Obtaining the slow and constant supply of slurry needed by a digester can be difficult on existing farms. If new farm buildings are to be constructd then the slurry system can be planned with digester operation in view. Settlement of solids in slurries is likely and while this may not change the volume of slurry it will change gas production per unit of slurry. Supernatants of pig slurries left in channels for some time have been found with up to about 9000 mg acid per litre instead of the usual 2–3000 found in fresh slurry. A rapid production of acids in slurry storage tanks may lower the pH quickly enough to prevent build-up of a methanogenic flora. In this case H_2 and CO_2 are produced as well as acids, and the methane-forming potential of these gases is lost. It is also possible, depending on the temperature and conditions within the slurry storage, that fermentation will be slow and a complete digester flora will build up which will convert settled solids to acids, H_2 and CO_2 and also convert these intermediates to methane and CO_2. Thus, when the slurry solids are finally cleared out from the channels and storage almost all the biogas-producing capacity has been lost and the slurry is useless so far as a productive digester is concerned. Apart from loss of biogas potential, storage of slurry solids can be dangerous.

Hydrogen sulphide will also be produced and a potentially explosive, suffocating and poisonous gas mixture can accumulate in the tank headspace. Fatalities to men and animals have been caused by gas produced from stored slurry. If solids are likely to settle then the slurry system may have to be changed to include scrapers or tank agitators to keep the solids suspended. The systems should be reliable, but as simple as possible to cut down capital and running costs as well as running energy. If a better slurry-handling system is installed its running costs should be debited against extra energy produced by the digester.

The third factor is the local climate and economics. In warm countries climatic heating is often sufficient for a mesophilic digester. Labour for loading and operating the digester is inexpensive and construction of small digesters can be simple compared to large, automated, digesters. In this type of location digesters can be run on small farms and virtually without external energy. India and China have the largest number of digesters working on combined farm and human wastes. The gas energy produced by such small digesters may economically run a boiler or provide gas for cooking and lighting. In this case three or four cows or ten pigs can make the digester viable and provide the energy needs of a small farm.

In Europe and USA where there is a high degree of automation and greater needs for energy the minimum digester size is larger. The smallest size is 2–300 pigs or about 50 cows. A digester of this size could provide useful heat energy from a boiler. To generate electricity about 2000 pigs or about 1–200 cows are required. These sizes may be complicated by the fact that on most dairy units cows are housed for only about six months of the year, for the remainder they are out grazing. Thus, usable excreta are produced for only about half the year. This is severely limiting economically unless alternative sources of feedstock can be found. A further factor is the use of the energy. More than a few hours gas production cannot readily be stored, so the gas must be used as it is produced. This may mean provision of alternative uses for different days or seasons, otherwise the gas will have to be wasted by flaring.

3.5. Types of digester

In Europe and USA the first research on farm wastes followed the sewage treatment practice and used stirred-tank, continuous-flow, digesters. Traditional water-industry construction techniques of large concrete tanks were too expensive for farms and low-cost prefabricated tanks, similar to those used for grain storage, were adapted for waste treatment. These subsequently proved to be also very useful for sewage-sludge digestion, especially for small communities (Noone and Boyd, 1980; see Chapter 4).

There are three basic types of digester:

(1) batch digesters;
(2) continuous-flow digesters for sludge digestion;
(3) retained-biomass or 'second-generation' digesters.

In recent years there has been much research on the use of the so-called 'second-generation' digesters, or retained-biomass reactors. These digesters cannot treat thick sludges with high concentrations of microbiologically-resistant suspended solids; they can only be used on wastewaters or sludges from which most of the solids have been removed.

A large number of small stirred-tank digesters have been built in developing countries for very small farms, particularly in India and China. Various designs were described by Horton (1980); the digesters described here will be mainly those suitable for large farms in Europe and similar areas.

3.5.1 Batch digesters

In batch digestion all the waste is introduced into the digester at once, together with an inoculum of active anaerobic bacteria. The mass is kept under suitable conditions and the bacteria grow and digest the feedstock. Gas production starts slowly, increases and then decreases and finally ceases. The digester is then emptied except for a small amount of material to act as inoculum for the next batch of feedstock and the process is repeated. In practice continuous-flow digesters have advantages over batch digesters, mainly because wastes are usually produced continuously and biogas energy is needed continuously. The only way in which batch digestion can deal with a continuous flow of waste or provide a continuous supply of energy is to have a number of digesters working out of phase. One will be loaded while another is supplying gas, and so on. Batch digestion is also more susceptible to toxicity; animal excreta often contain a high concentration of ammonia which can inhibit digestion of the undiluted single charge of waste. However, batch digesters can have some advantages with high solids wastes such as poultry litter or excreta plus straw or other bedding, or for crop wastes. Batch digestion avoids the need for continuously introducing and removing feedstock and the actual operations can be done with tractor shovels or similar equipment. The digestion of crop wastes such as corn stover and straws has to be done as a batch process unless the material can be made into a slurry with other wastes. Jewell (1980) suggested the use of very large batch digesters with excreta (as bacterial inoculum) plus straw or other vegetation mixed to give a solids content of 30–40%. These digesters could produce gas for a year or so on one charge and it seems possible that the microbial activity could provide enough heat to keep a lagged digester at mesophilic or even thermophilic operating temperatures. So far as is known full scale large batch digesters have not yet been built. One prototype for 'dry' manure on a French farm was described by Demuynck et al. (1984). It was a batch digester for cattle waste and straw bedding. Clausen and Gaddy (1983) described a batch digester plant with four 20 m^3 digesters operating on a 10% slurry of milled grasses and hay. Operation was on a 60 day cycle at 35°C, so that overall gas production was continuous. The Chinese have built a number of small-farm batch digesters for crop stalks and pig waste of 25–30% TS. These are hand-loaded and operate at ambient temperature over a cycle of about four months in summer and five months in winter (Guo-Chao Sun et al., 1987; Ke-Xin Liu et al., 1987). An 80 m^3 batch digester running on a 27% TS mixture of cattle manure and crop wastes was described by Molnar and Bartha (1989). This was a concrete bed with a plastic cover which acted as a gas collector. A novel point was that the feedstock was composted under forced aeration for 2–3 days and this raised the temperature of the material to 55°C. Aeration was stopped and anaerobic digestion was allowed to start. With no further heating, digestion at a gradually decreasing temperature took place over the next 30 days. The digester was then emptied and the process repeated. A number of these digester units were built in pairs to give a continuous digestion and gas production. A municipal landfill site with gas recovery is a very large batch digester which can continue to produce gas for years. There are now some digesters for the continuous digestion of the solid organic matter from municipal garbage, but these have not been adapted for farm use; indeed there are only a few in use for municipal waste at the moment.

In fermentation technology a variant on the batch fermentor is the 'fed-batch'. In this system the fermentation is allowed to proceed in a partly filled vessel. As the fermentation nears a conclusion more medium and substrate is added. The added substrate then ferments and the addition is repeated. When the vessel is full the contents are emptied, except for a residue as inoculum for the next batch, and the process is repeated.

This type of reactor has been used in anaerobic digestion of farm wastes, although here the reactor is filled more or less continuously in response to the excretion of the animals and not at intervals determined by the progress of the fermentation. It has been called 'continuously expanding' digestion by Hill et al. (1981) and Hill and Bolte (1986). A special tank may be built, but, by suitable sealing, a slurry-storage tank under a solid-floor animal house may be made gastight and become a digester. The excreta are collected in channels behind the animals and run from there through liquid-seal inlets to the tank, which fills up with waste over many months. Slatted floors may also be used to feed the digester. There are examples of these systems in use (Wellinger and Kaufmann, 1982; Demuynck et al., 1984). The digesters are usually run unmixed and without external heating, but are insulated to retain the warmth from the fresh excreta and the animal house. Nevertheless, some form of heating may be desirable as at temperatures below about 18°C start-up of digestion can be difficult without a large inoculum of bacteria adapted to low temperatures, and the rate of gas production at 10°C (which can be the temperature of some under-floor digesters in winter) is too low to be commercially useful. A temperature of 15°C has been recommended as the lowest temperature for running under-floor digesters and start-up can be expedited by temporarily increasing the temperature to 25°C (Wellinger and Kaufmann, 1982; Cullimore et al., 1985; Zeeman et al., 1988).

Some of the digesters are mixed by externally-driven propellers, but in some systems (e.g. Haga et al., 1979) submersible pumps provide both mixing and heating. However, in the author's experience centrifugal-type pumps, submersible or non-submersible, can be subject to rapid wear from the grit contained in farm slurries. The digester is often finally emptied at the end of the winter season when cattle are turned out to graze, and the digestion can be re-started at the beginning of the next season. This is one of the advantages of the anaerobic digestion, particularly sludge digesters; digesters can be left unheated and unloaded for long periods and started up again, without reinoculation. Start-up is by heating and adding fresh feed at a low rate and then building up to the normal loading rate in a comparatively short time. However, some fed-batch under-floor systems are run in continually-occupied animal houses and the tank is emptied when the waste is required as fertilizer. An under-floor tank can be run as a fed-batch until the tank is filled and then as a continuous-flow digester with waste overflowing as fresh excreta pass into the tank (e.g. Sutter and Wellinger, 1988). Retention times in such systems are usually fairly arbitrary as mixing can never be perfect. In most cases retention times should be long. This low-level tank system could be a relatively cheap way of adding a digester to an animal house by covering already-existing under-floor tanks or channels.

A further type of batch digester, so far as is known tested only on a laboratory or pilot-plant scale, uses a vertical cylinder. The dry, but permeable, waste (e.g. straw) is introduced, with an inoculum if necessary, into the middle of the cylinder resting on a grid. Water percolates down through the mass and is pumped from the bottom to the top to recirculate. Operation at ambient temperature or 35°C is envisaged. Hall et al. (1985) operated small digesters on this principle at a cycle time of 40–70 days, at 35°C, with mixtures of straw and cattle waste of 26% TS. The circulating liquid helps to distribute the bacteria evenly through the solids and to prevent unreactive volumes, or pockets with unbalanced reactions, developing. The biogas from a batch reactor can be used in the same ways as that from a stirred-tank or other form of digester.

3.5.2 *The stirred-tank, continuous-flow, digester*

The stirred-tank is the basis of most farm digesters. Fresh waste, as a slurry usually, is added continuously or at short intervals and the same volume of digested waste overflows. The anaerobic deep lagoon is a stirred-tank in its simplest form, once it is full. During initial operation it acts like a fed-batch digester. Waste flows in and is stirred by gassing and thermal convection currents, the solids usually settle and the retention times of solids and liquids can be quite different. It is possible to cover the lagoon and collect the biogas for use in engines or boilers (Balsari and Bozza, 1988; Safley and Westerman, 1989). Oleszkiewicz and Koziarski (1986) have discussed the kinetics of digestion in lagoons and Cullimore *et al.* (1985) have studied the effects of temperature. Obviously, in temperate climates the rate of digestion will be slow overall and may be almost negligible in winter (Balsari and Bozza, 1988). But the covering should obviate one of the problems of lagoons, a tendency for evaporation in hot weather to dry up the slurry and for rain to dilute the slurry. However, for optimum gas production and pollution reduction a more controlled system is needed. Thus, the typical stirred-tank digesters are above- or partly below-ground tanks fitted with pumps for controlling flows, mechanical mixing and heating, with temperature control.

Many full-scale stirred-tank digesters have been built on farms in Europe and elsewhere. The main ones in Europe have been described by Demuynck *et al.* (1984). Not all the digesters built a few years ago are still working, some were built partly as experimental projects, some have been dismantled for different reasons. The size has varied and construction has been by the farmer himself or by companies specializing in digesters. The materials and method of construction have thus varied. Most of the digesters are above-ground tanks, usually cylindrical, but some are underground tanks. The method of loading is usually by a pump, although those below ground can be gravity loaded by slurry flowing down channels from the animal house. Although all can be referred to as 'continuous-flow' digesters, the loading is generally intermittent with slurry being added every hour or once or twice a day. There are various types of mixer. Different types of paddle and Archimedean-screw mixers are used rotating freely or in draft tubes. Most modern tanks are now mixed by recirculating some of the biogas to the bottom of the digester tank from where it can rise to the top either as free bubbles or as bubbles confined in draft tubes. Mixing is usually intermittent at intervals of about an hour. Heating is generally by hot water in simple heat-exchangers as pipes or plates immersed in the tank. The hot water can come from a gas boiler or from an engine running on the biogas. Engines can also generate electricity to run digester pumps. The ancillary equipment depends on the use (if any) of the biogas. Boilers and engines are the most usual users of gas, but gas from a big plant could be used in grain driers or other farm equipment. The excess heat and electricity produced from the digesters can be used in the farmhouse or in the animal house for heating, lighting and ventilation.

The usual stirred-tank type of farm digester is used for treating pig or cattle faecal slurries of about 5–14% TS. Digesters have also been used for poultry excreta, but as these are more or less solid as produced they have to be slurried with water before digestion. Alternatively the waste-collection system in the poultry houses must be a water-flushing type which slurries the waste as it is collected. Mixtures of the different wastes have also been used. There have been a number of papers describing experiences with farm stirred-tank digesters. Summers *et al.* (1984) for example describe a 680 m^3 digester running on piggery waste with a boiler and gas-engine generator. Safley *et al.* (1987) describe experiences with a digester running on poultry waste and Poels *et al.* (1983) describe integration of a digester with a pig farm. Hobson and Feilden (1982) have reported on some aspects of the construction and use of farm digesters based on experience with the 680 m^3 digester mentioned above and with two smaller ones running on pig and cattle wastes. A 100 m^3 digester of novel design has been running for many years at a farm research institute in Germany (Baader, 1985). The tank is completely filled with the digesting slurry which is mixed by a screw-mixer in a draft tube. The digester liquid is circulated from bottom to part-way up via an external heat-exchanger where feed is added to the recirculating liquid. Slurry feeds can be fed by a pump, but an auger-type pump can also feed solids into the tank along with the recirculating liquid. The plant has been used to digest cattle-waste slurries and vegetable wastes of various solids concentrations. The gas runs a gas-engine generator.

3.5.3 *The tubular or plug-flow digester*

The tubular fermentor is not mixed and the substrate is introduced at one end of a horizontal tube and flows as a plug through the tube. The plug of material is inoculated as it enters the tube and the course of fermentation is equivalent to that in a batch digester as the plug of material progresses through the tube. If the retention time of the plug in the tube is long enough, then, like a batch fermentation, effectively all the substrate can be used up. Thus, in a theoretical tubular digester the effluent should contain much lower concentrations of volatile fatty acids and other degradable substances than that from a stirred-tank digester running at normal retention times (Hobson *et al.*, 1981). With animal excreta the inoculum bacteria are in the manure and so no additional inoculum should be needed, but an inoculum can be obtained by recycling some of the effluent liquid to the feed. In practice however, friction from the walls and the effects of gas production mix the contents of the tubular digester. The convection currents from the heating system add to the mixing. Solids from slurries also tend to settle out. The tubular digester therefore becomes a horizontal stirred-tank with less than optimum mixing and different retention times for solids and liquids. The settling of solids can be responsible for the apparently better degradation reported to take place in some tubular digesters than in stirred-tanks of the same nominal retention times. The solids have a longer retention time than in the better-mixed tank because of settlement.

One of the early types of tubular digester was a long trough in the ground, partly filled with slurry (Hayes *et al.*, 1980). Weirs acted as gas seals and allowed the slurry to flow in and out either by pump or by gravity. The trough had a flexible cover which acted as a gas-holder. In temperate climates the digester can be heated with hot-water pipes, but without heating, this digester type can be a cheap and easy system for use on small farms in hot countries (often known as 'red-mud plastic' digesters). This type of construction is not suitable for large digesters. Square-section tubes have also been used (Fry, 1974), and there are commercial digesters on a much larger scale made from glass-fibre plastics (Fig. 3.1) which have a tubular construction. However, these and some other horizontal tanks are not designed as plug-flow digesters but as stirred-tanks.

One of the problems of the horizontal digester is scum formation on the surface of the liquid. A manually operated scum rake was used by Fry (1974) and a pilot-plant tubular digester described by Field *et al.* (1985) had a hand operated liquid–surface agitator. Some horizontal digesters have a series of paddles or a screw rotating at low speed about a longitudinal axis as a stirrer (one described by Zeeman *et al.* (1988) is typical). Some digesters with an unstirred tube inclined at a few degrees to the horizontal have been tested. A small prototype plant showed that solids were retained, but that gas production was similar to a stirred-tank digester. Some mechanical problems were also apparent: e.g. blockages in the overflow (Floyd and Hawkes, 1986; Chapman *et al.*, 1988).

3.5.4 *Retained-biomass digesters*

The previous types of digester are used for digestion of solids or of sludges containing biodegradable solids plus degradable material (e.g.

Fig. 3.1 An award-winning farm-waste treatment plant showing the digester installed in a pit for easy loading. The butyl-rubber bag in the background is for gas storage. Photograph courtesy of Farm Gas Ltd, Bishop's Castle, Shropshire, UK.

acids) in solution. The retention time measured in days is set by the time needed for degradation of the solids to acids, hydrogen and carbon dioxide. These retention times are sufficient for growth of the slowly growing methanogenic bacteria on these fermentation products. If the feedstock contains only dissolved sugars or similarly easily-fermentable material then the conversion of these to acids is rapid and the bacteria involved can double in numbers in a matter of a few hours. These acid-forming bacteria could keep up their numbers in a purely fermentative culture in which the feedstock flow gave a retention time of a few hours. If the retention time is less than the minimum doubling time of the bacteria then they will be washed out of a stirred-tank fermenter faster than they can grow and the system will fail. The growth rates of the acid-utilizing methanogenic bacteria are low and this sets a minimum retention time of three or four days for acid metabolism in a stirred-tank digester. For complete conversion of an easily-fermented feedstock into methane the retention time is then governed by the growth rates of the methanogens. For some feedstocks with not too high flow rates the problem of the difference in growth rates can be solved by the 'two-phase' digester which will be described later. For factory wastewaters flowing at much higher rates than farm wastes or sewage sludges a liquid retention time of three or four days would result in a very large stirred-tank digester and the gas production per unit volume of digester would be low. However, if an active mass of bacteria, fermentative and methanogenic, can be retained in the digester for sufficient time for the bacteria to grow (days) the wastewater feed can flow over this mass at a high rate (retention time, hours) while dissolved pollutants are completely absorbed and converted to methane by the retained bacteria.

There are three types of retained-biomass digesters:

(1) immobilized biomass; filter and fluidized-bed digesters;
(2) flocculated biomass; sludge-blanket digesters;
(3) recycled-biomass; contact digesters.

These are discussed in detail in Chapter 5: they have been applied to farm wastes and wastes from factories processing agricultural products. For farm wastes the feedstocks are either liquids such as silage effluent which are formed with little suspended solids, or animal-house slurries from which most of the suspended solids have been removed by a mechanical separator or by settlement in a lagoon or storage tank. They might also be used as part of a secondary treatment system for final purification of a primary digester effluent.

(a) *The anaerobic filter.* In this system the anaerobic bacteria are immobilized on the surface of an inert support medium which more or less fills a tank. The support can be irregularly-positioned, centimetre-size, lumps of stone (Young and McCarty, 1969), brick or plastic or regularly-arranged tubes of unglazed pottery or similar material (van den Berg and Lentz, 1979). The bacteria are retained by growth attached to the support and by entrapment in the interstitial space. The retention of the bacteria in the tank is not infinite and some are continually sloughing off the matrix, but overall their retention time is between days and weeks. Suitable baffles or a multiple inlet system is used to ensure even upward or downward flow of the feed liquid through the support matrix. The biogas escapes through a separate port at the top of the filter. Dilute wastes may be treated in retention times of hours rather than days. Filters have been used for silage liquid (Barry and Colleran, 1982) and the supernatant from settled pig waste (Kennedy and van den Berg, 1982) for instance. They could be used for milking-parlour wastewaters and for similar liquids.

The anaerobic filter can be blocked by accumulation of small particles from the feedstock and with some filters used in factories arrangements are made for back-washing the filter at intervals. A development of the filter, which is less likely to block, and which presents a much bigger surface for growth of the bacteria and contact with the feedstock, is the fluidized-bed digester. Here, the bacterial support matrix is small (1–5 mm) granules of sand, plastic or glass. The mass of the support material at the bottom of the filter is expanded and fluidized by the upward flow of the feed liquid combined with a recirculation of some of the effluent to increase the upward velocity of the liquid. Jewell (1980) was one of the originators of this type of digester for farm wastes and, like the filter, it can be used for slurry supernatants or wastewaters from farm processes. These types of digester can be run at mesophilic temperatures around 30°C, but are more generally run at ambient temperatures. The filter is obviously the mechanically simpler of the two.

(b) *The flocculated-biomass digester.* An alternative method of bacterial retention is to encourage the bacterial consortia to form a high-density granular or flocculant sludge which can be kept in the reactor vessel by a balance between the upward flow of the feedstock and gravity settling. The most common design is the Upflow Anaerobic Sludge Blanket (UASB; Lettinga *et al.*, 1979). The biogas escapes through the top of the reactor while any bacterial flocs that rise with the gas are retained by baffles. The principal use of the UASB digester has been in treating wastes from factories processing fruits and vegetables (see Chapter 5). On the whole the filter is a cheaper and easier-run plant for farm wastes.

(c) *The contact digester.* In this type of digester the bacteria-containing suspended solids in the effluent from a stirred-tank digester are separated from the effluent and recycled back to the digester. This increases their effective retention time. Separation of the bacteria does present problems; the gas can disrupt settling bacteria. Centrifuges, filters, baffled tanks, with a vacuum to remove gas, or cooling or chemically-assisted flocculation have all been used.

Again, this type of digester is better fitted for use in a food-processing or similar factory than on a farm.

(d) *Two-stage, two-phase and hybrid digesters.* With the stirred-tank ('chemostat') fermenter in which one bacterial species carry out one particular reaction, better utilization of substrate and lower retention times can be obtained by using two or more tanks in series. A large number of tanks in series approximates to the theoretical tubular fermenter. In practice more than two tanks cannot usually be justified. Hobson *et al.* (1984) tested pilot-plants with two tanks in series, with different volumes and so different retention times, treating piggery slurries. They found, as predicted, that the second tank reduced the levels of residual acids in the effluent from the first tank so that an effluent of lower BOD could be obtained from two tanks running at lower overall retention time than one tank. However, digestion of the solids was the same at the same retention time in one tank or in two. While the dissolved bacterial substrates behaved as the theory of two-stage cultures predicts (Hobson *et al.*, 1984), the degradation of fibre particles by bacteria colonizing their surfaces is a function of retention time whether this retention time is spent in one tank or in two or more. This type of two-stage digester could then be used as a form of primary and secondary treatment for sludges, with a second digester much smaller than the first and so of shorter retention time. A few farm digesters have been built with two tanks in series, but there has been no theoretical basis for the design. The use of two tanks obviates separation of solids before secondary treatment, but if solids are allowed to settle or are mechanically separated then a filter or other retained-biomass digester can be used to reduce concentrations of residual acids (and so BOD) in a digester effluent. The filter digester of Kennedy and van den Berg (1982) was suggested for use with a feedstock of low-solids pig waste from a lagoon or storage-tank supernatant. In this case, as well as allowing some settling the lagoon would act as a low-rate, first-stage digester. Oleszkiewicz (1983, 1985) conducted experiments on systems with a primary stirred-tank digester followed by a filter, with final aerobic polishing for complete treatment.

The 'hybrid' types of digester are two-stage types in which slurries can be digested. One example was used by Callander and Barford (1984) at a pilot-plant scale. In these experiments a 4% TS pig slurry was pumped into the bottom of a tower. The solids remained near the bottom and were slowly digested (residual solids were removed at intervals) and the liquid then passed upward through a sludge-bed where dissolved substrates were digested. Hybrid digesters acting on the same principle, but with a filter half-way up the tower have also been described, but, so far as is known, have not been used on a large scale.

'Two-phase' digestion is not the same as 'two-stage'. In the two-phase digester an attempt is made to separate the fermentative and methanogenic steps of digestion so that each may be carried on under more optimized conditions than in a tank in which all the reactions occur.

The only practical way in which the reactions can be separated is by using retention time. Fermentation of dissolved sugars and similar substrates can be carried out in a fermenter with a retention time of a few hours, as the bacteria grow on these substrates with a doubling time which allows a fast flow through the fermenter tank. The acid-utilizing methanogens, however, have growth rates which require retention times of at least three or four days. Thus, if two tanks are connected in series with the first being of smaller volume than the second then a suitable substrate can be acidified in the first tank without growth of methanogens, and the acids can be converted to methane in the second tank which has a long retention time. Since fermentation can take place at a lower pH than methanogenesis (c. 5–6 as against 7 or over) further selection for fermentative bacteria can be obtained by allowing the pH to fall in the first tank (Cohen et al., 1979). Thus, food-factory wastes, high in sugars and low in solids, can be digested efficiently in the two-phase system, although some potential biogas is lost unless hydrogen and carbon dioxide given off from the fermentation are channelled to the methanogenic reactor. It is also possible to digest easily-degradable sewage sludges in a similar two-phase system (Ghosh, 1984). However, this approach does not work properly with wastes containing large amounts of lignified fibres, resistant to hydrolysis, such as farm wastes. Hydrolysis and fermentation of the solids in animal wastes requires a long retention time and so this allows methanogens to develop. The pH optima for cellulolysis and methanogenesis are similar, so it is impossible to completely separate the reactions by pH control, although the cellulolysis and fermentation can take place at about pH 6·5, at which pH methanogenesis is inhibited. Colleran et al. (1982) suggested a simple two-phase process in which the slurry stored in animal-house tanks was allowed to ferment and go slightly acid at ambient temperature. The acidified supernatant was then pumped to a second-phase anaerobic filter where the acids were converted to biogas. In this system the solids fermentation is slow and incomplete and methane and carbon dioxide are lost from the first tank. This may not be a serious drawback for a small farm where a simple plant is the main consideration, but it could be so on a large farm where a return in power generation was a main part of the digester plant. A variation on this scheme is to use the percolating digester previously mentioned but allow the solids and percolating water to go slightly acid and the solids to ferment. The acidified percolate is then passed to a methanogenic filter or sludge-blanket type of digester ('methanizer') where acids are converted to methane. (The pH of the working methanizer settles to about 7 or rather higher.) The effluent from the 'methanizer' can be passed on, or returned to the 'hydrolyser' tank for further recirculation. A number of experimental plants have been made on these lines, mainly for digesting crop wastes, but as far as is known no full-scale farm plants have been made (e.g. Viturtia *et al.*, 1989 (fruit and vegetable wastes); Coble and Egg, 1987 (sorghum)). Such plants can be run with a number of batch hydrolysers connected in sequence to a methanizer. However, it has been found that towards the end of the hydrolysis, when the rate slows and the pH rises as acid concentration becomes low, the methanogenic bacteria begin to grow in the hydrolyser.

3.6 Digester operation

There are four important factors in the process:

(1) Retention time (RT) and mixing,
(2) solids concentration in the feed and in the digester,
(3) temperature,
(4) inhibition.

3.6.1 *Retention time*

Small digesters, pilot- or laboratory-scale, can be mixed relatively easily and so in the stirred-tank digester liquid and solids have the same retention time. Fibrous solids from cattle and pig waste tend to float or sink in the digester liquid, and mixing in large digesters may not be sufficient to prevent this happening. Floating solids cause blockages in overflow weirs and have been known to dry out and completely cover the top of the liquid, impeding gas escape. One reason for the design of the completely-filled digester mentioned earlier was to overcome problems of surface scums of floating solids. Settled solids remain in the digester for longer than the liquid retention time and also may become dead areas where no reactions take place. The formation of sediments at the bottom of the digester is aided by grit in the feedstock. The inorganic content of slurry solids from animals kept indoors is about 20–30%. This is made up of salts from the excreta and grit abraided from the animal-house floors. Dairy-cow stalls may be covered with sand to reduce the problems of udder infections associated with wet straw or sawdust beddings. Grit, sand and stones can be picked up from earth-floored feedlots. The grit can cause problems with blockages and abrasion of pumps and pipes and can also settle in the digester along with fibres. Settled material like this builds up in unmixed 'corners' of the tank and can reduce the effective volume of the tank. The retention time RT can then no longer

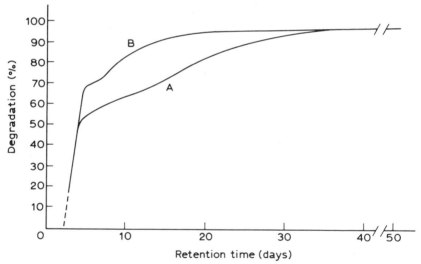

Fig. 3.2 The effect of the retention time on the percentage degradation of the biodegradable portions of the solids in cattle- (A) and pig- (B) waste slurries at about 35°C. These are generalized graphs based on data such as those given by Hobson et al. (1981).

be related to the empty volume of the tank and so the feed to a tank digester may then have an actual RT quite different from the design RT. As the RT will be shorter than the design then the gas production and the pollutant removal will be lower than the calculated values. Digesters are often fitted with a conical floor sloping to a large-bore valve through which grit-laden slurry can be periodically removed.

Figure 3.2 shows that a retention time of 10 or 12–30 days is usual for tank digesters treating faecal slurries. At the retention times used in practice, the digestion of solids and acids is not quite complete. The course of digestion is asymptotic as RT increases. An RT has to be chosen which is a compromise between maximum digestion and a reasonable-size tank. At a given flow of waste increasing the RT means enlarging the size of the digester. Eventually only a few percent more digestion may be obtained with orders of magnitude changes in tank size. Similar considerations apply to the extent of pollutant removal with RT, and other types of digester.

3.6.2 Solids concentration

For a digester operating at near to optimum retention time the gas production will depend on the solids concentration of farm slurries, whether the digester be a single-stage, a two-stage, or phased system. There is, however, a maximum solids content which can be effectively treated if slurries are to be pumped and piped into the digester: above about 8% TS mixing and pumping slurries becomes very difficult.

Fresh cattle slurries are about 12–14% TS. Such slurries can be handled by scrapers in channels and gravity-fed to a digester. Large-bore Archimedean-screw, auger-type pumps, such as are used for moving grains can move thick slurries, but cannot usually move them in long lengths of associated pipelines. In general, it is the larger particles in slurries which are the most difficult to move and these are also the least biodegradable. Separate removal of these large, usually fibrous, solids makes both slurry handling and slurry digestion easier without much loss of the theoretical gas production from the whole slurry (Lo *et al.*, 1983; Pain *et al.*, 1984; Summers *et al.*, 1987). This type of material can be separated by mechanical means and the large particles then form a stackable solid that can be composted for use as a soil conditioner, or even as an animal bedding. On a large scale the better handling properties of the slurry and the smaller digester needed (for a shorter RT since the more resistant solids have been removed) have to be balanced against the capital and running costs of the separator. Such systems can be designed to remove various sizes of solids. If most of the solids are removed, leaving only a liquid of 1 or 2% solids, then this liquid may be treated in a retained-biomass digester, but removal of only the largest solids and use of a stirred-tank digester has generally been the practice. Poultry wastes may have a solids content of 20% or more, particularly when mixed with shavings or other litter. These have to be diluted with other wastes or water to make digestion in a conventional stirred-tank possible. Poultry wastes might be digested as solids in the batch digester designs, previously mentioned, but digestion of very-high solids wastes may be inhibited by excessive production of toxic metabolites, principally ammonia (Hobson et al., 1980; Robbins et al., 1989). There is also the possibility that inhibitory antibiotics, as well as disinfectants, etc., may be concentrated in solid wastes. Dilution of the solid wastes, as mentioned, may reduce toxic compounds to non-inhibitory concentrations.

3.6.3 Temperature

Anaerobic digestion will take place at usable rates across a broad temperature range of 15–65°C. The mesophilic range, from about 25–40°C is generally considered as optimum for heated digesters. Table 3.3 shows some effects of temperature in the mesophilic range. It is the rate rather than the extent of the reactions that is affected by temperature, and a lower digester temperature can be compensated

Table 3.3 The effects of temperature on anaerobic digestion of piggery wastes in stirred-tank, continuous-flow, digesters

	Temperature (°C)					
	15^a	25	30	35	40	44
Gas (litres/kg TS)	0	260	300	300	360	420
Solids degradation (%)	0	—	33	36	37	38

[a] From van Velsen (1977), other results from Summers and Bousfield (1980). Ten–twelve day RT.

for by a longer retention time (Chen, 1983). The data in Table 3.3 were obtained from a stirred-tank digester with a 10–12 day RT. Most of the small climatically heated Indian and Chinese digesters do run at less than 25°C for much of the time, but with a retention time of about 50 days. Similar amounts of gas per cow can be obtained as from a digester running at 35°C and 20 day RT.

Gas production from a mesophilic digester will fall away rapidly if the temperature is raised over 45°C. Eventually, however, if the temperature is raised still further another group of bacteria can become established; the 'thermophilic' bacteria. The thermophilic range is narrow, 50–65°C. Much work has been done on thermophilic digestion on a small scale, particularly with cattle wastes. This is because the warmer temperatures accelerate the hydrolysis of the higher solids content present in cattle waste than in piggery wastes (Varel et al., 1977, 1980; van Velsen et al., 1979; Hashimoto, 1983; Hill et al., 1985). Thermophilic digestion is also better for the destruction of pathogenic organisms in slurries than is mesophilic digestion. Pathogen removal is a function of temperature and time (Sanders et al., 1979; Turner et al., 1983; Plym-Forshell, 1983; Hobson, 1984; Olsen et al., 1985; McKain and Hobson, 1987). There are, however, problems with thermophilic digestion. Toxicity and inhibition of the process frequently occurs, caused by ammonia or carboxylic acids (van Velsen, 1979; Zeeman et al., 1985). Thermophilic digestion of pig waste has generally been unsuccessful, probably mainly due to the ammonia content of the waste. There are also problems with the residues after digestion. Solids settlement after thermophilic digestion is not as good as after mesophilic digestion and the treated material can be very malodorous.

Thermophilic processes have been mostly suggested for energy production rather than pollution control. More heat is required to run a digester in the thermophilic range than in the mesophilic, unless the feedstock is a factory wastewater produced at a high temperature. However, the retention time is shorter at thermophilic temperatures so gas production per digester volume is greater. Although the digester is smaller than for mesophilic digestion at the same slurry flow, the greatest heat input to a digester is heating up the feedstock and the heat required to raise the feedstock from 10 to 60°C is twice that needed to raise the feed to 35°C. The fact that farm feedstocks are cold, that slurry and pipes at 60°C are unpleasantly hot to touch, and the extra capital costs involved in building a thermophilic digester means that they are rare. Demuynck *et al.* (1984), in their survey of agricultural biogas plants in Europe found that 5% of digesters (mainly under-house type) were operating at ambient temperature, 12% were working at 25–30°C, 64% at 31–35°C, 16% at 35–40°C and only 4% were thermophilic, working at higher than 50°C.

3.6.4 Inhibitory materials

Farm-waste, stirred-tank, digesters are usually extremely stable in operation and will tolerate without difficulty large variations in loading rate, periods of no loading and heating or slow changes in temperature. Rapid changes in temperature or overheating are not so well tolerated, but these are unlikely because of the large heat capacity of a big tank of waste. The heating capacity of the heat-exchange system is also usually such that even if the water overheats the temperature rise in the digester will be slow.

Failure of digestion because of mechanical breakdowns is unlikely, but inhibitory substances may cause problems. Ammonia toxicity can be a problem, but sudden increases in ammonia concentration in the feedstock are unusual. These are mostly caused by taking a batch of feed slurry that has stood for some time in the animal house. Even then the feed would be diluted within the digester and it would take 10 or 20 day's feeding to approach the replacement of the normal digester contents with toxic feed. One particular batch of high-ammonia slurry should therefore not affect a well-operating digester. The same reasoning applies to other toxic contaminants in the feedstock. In the case of ammonia there is much evidence that the bacteria can adapt to what, in the short term, would be inhibitory concentrations of ammonia. Digesters have been successfully run with slurries containing much higher concentrations of ammonia than those inhibitory to unadapted bacteria (*c.* 1800 mg/ml). Antibiotics, and biocides used in cleaning and sterilizing compounds may cause problems. The bactericidal effects of these compounds are a function of time of exposure and concentration. The large-scale addition of antibiotics to animal feeds to improve weight gains is now less common than it was. If the antibiotics are used in small amounts as feed additives to the young animals in a large herd or therapeutically for just a few animals, then any antibiotic residues in the excreta will be diluted to below the toxicity threshold by the excreta from the untreated remainder of the herd. This was found for the digester described by Summers *et al.* (1984). Some other feed additives are now allowed (e.g. monensin, lasalocid) and these may be inhibitory. Further work is required on such additives. However, the bacteria can adapt to many of the feed-additive compounds and inhibited digestion is reversed after some weeks. Disinfectants used for cleaning and sterilizing animal houses or milking parlours are also a potential cause of digester malfunction if they are used indiscriminately and get into the digester feedstock. These compounds may also be diluted by the main volumes of waste and not be inhibitory. Any source of antimicrobial agents must be considered in designing a digester. The toxic compound can then be diverted or can be deliberately diluted to prevent problems. Experiments on the concentration, exposure time and consequent inhibitory effects of various biocides have been reported: for instance by Varel and Hashimoto (1981); Hilbert *et al.* (1984); Poels *et al.* (1984). Sulphide can be toxic to digester bacteria, but this is not usually a problem with farm wastes. Problems can arise with some wastewaters from processing of farm crops in factories (see Chapter 5).

3.6.5 Start-up

Start-up of farm digesters is relatively easy. The slurry is well buffered and 'souring' of digesters is comparatively rare. Animal excreta contain the bacteria needed in a digester, although in low numbers. So a digestion can be developed by initially operating the digester at low loading rates and long retention times. The slurry input is then gradually increased to the working rate as gas production develops. This may take up to eight or nine weeks (Hobson and Shaw, 1974). Quicker start-up can be obtained by partly filling the digester with sludge from a working digester or with slurry that has been stored for some time and has already begun to digest. A beneficial extra inoculum can be obtained by recycling some effluent with the feed. Commercial digester suppliers will include start-up in their building of a digester. Safety aspects of start-up and running of digesters are covered in some handbooks such as the one produced by BABA (1982).

3.7 Digester construction

Many of the earlier farm digesters were partly experimental and used a wide variety of materials and designs. Digesters have now become less varied and standard designs built by a few contractors are now available. These are built to different sizes, but often larger capacities are derived by constructing two or more units in parallel. The survey by Demuynck *et al.* (1984) gives details of construction of the digesters surveyed and some notes on construction in general; the paper by Hobson and Feilden (1982) also gives some notes on construction as do the papers previously quoted on running of farm digesters.

Most large digesters for intensive farm units are above-ground vertical, cylindrical tanks and the usual materials of construction are steel or concrete. Glass-reinforced plastic can be used for smaller digesters. Digesters can be made from mild steel, but a better material is the glass-enamelled steel used to construct silos, slurry tanks and other farm storage. These tanks are prefabricated in various sizes and are bolted together on site. They can be comparatively easily drilled to take pipes and other equipment. Concrete is self-insulating, but the steel tanks need to be insulated, inside or out, with foam materials covered with a layer of impervious paint or other coating. Digesters for small farms, in Europe but particularly in India and China, are frequently wholly or partly underground and can be of rectangular, spherical or cylindrical shape. The most common materials of construction are concrete, brick or stone lined with an impervious cement. Some European digesters have been constructed as horizontal steel tanks, but these have been of only a hundred or so cubic metres maximum size.

The small tubular digesters in use are mainly cement troughs in the ground covered with a plastic gas-holder. Early tubular designs in Africa were made from square-section concrete tubes placed end to end to produce long tanks. Gas was collected separately. Above about $200–300\,m^3$ building above ground is cheaper and simpler than building below-ground tubes. Some steel tubular digesters have been built above ground, but not so far as is known at more than pilot-plant scale. Retained-biomass digesters are tanks of similar construction to the other types, but are probably most generally made of steel.

It is difficult to produce a cement finish which is entirely impervious to liquid and gases, particularly where there are regular ground movements. The best solution seems to be regular relining of the tanks with a cement plaster. Many proprietary sealing paints are abraided and eroded by the slurry and do not tolerate the warm, wet aggressive atmosphere of the digester. A liquid polyurethane has been advocated. Any insulating coating on the outside of a digester must be sealed with a water-proof paint or plastic otherwise the insulation properties can be lost as the insulation gets saturated with rainwater.

3.8 Ancillary equipment

Once the anaerobic digester becomes larger than that associated with a small-holding then about 50% of the capital costs become associated with gas and slurry handling. Typical process flow diagrams are given by Demuynck *et al.* (1984), Summers *et al.* (1984), Poels *et al.* (1983), Safley *et al.* (1987) and Hobson and Feilden (1982) amongst those whose papers have already been quoted in other contexts.

3.8.1 *Gas storage*

The most economical gas storage and use is at low pressure, and this limits storage to between four and eight hours production. For some digester designs the gas storage acts merely as a buffer between production and use with only an hour or two production stored. Large gas-holders are expensive and can have a cost similar to that of the digester. Some storage can be provided in the head-space of the digester and this can be increased by provision of an expandable top to the digester. Integral gas-holders are common on the small digesters used in developing countries. Many of the large sewage digesters have integral gas-holders, but not many farm digesters have been so fitted. Small integral gas-holders can be rubber or plastic balloons covering the digester top. They usually have a system of weights to equalize pressures as the balloon fills and empties (the fabric does not expand to any extent and the gas inflates a fixed volume). At least one large (some hundreds of cubic metres) digester in Europe was fitted with such a gas-holder, but a better system is the metal roof fitted to sewage digesters and the large farm digester described by Summers *et al.* (1984). These roofs are sealed by immersion of the rim of the holder in the digester slurry or a separate water channel. Roller guides are provided to control the rise and fall of the holder.

Alternatively, and more commonly on a large scale, gas is collected in a separate holder. The separate holders are made in the same way as the integral holders, as balloons or as tanks sealed in a bottom water tank. Again, weights to control the filling of balloon holders, and guides for metal gas-bell holders, are needed. With all these holders the aim is to supply gas at as near as possible constant pressure of a few centimetres of water gauge. In most cases the gas may be used for cooking and in boilers and engines without purification. The analysis of the gas is from about 55–75% methane (the remainder carbon dioxide). Gas at 50% methane is combustible in boilers or engines (some figures on biogas composition and details of use in engines and boilers are given by Hobson and Feilden (1982)). Excess water vapour in the gas should be removed by condensation traps in the gas lines. Hydrogen sulphide in the gas may cause problems. This gas corrodes metals either as the sulphide or as oxidation products after combustion. However, farm digesters usually produce biogas with about 0·1–0·2% H_2S and this can be used in engines or boilers without serious difficulties. Engines need more frequent oil changes than do those running on pure gas and there may be some eventual valve wear, but purification of the gas involves extra equipment and attention which may, overall, cost more than any engine servicing. Some notes on engines and boilers are given by Hobson and Feilden in the paper previously quoted. Hydrogen sulphide can be removed from biogas by physical or chemical means, but some methods for gas purification are applicable only to larger volumes of gas at higher pressure than in farm digester plants. The easiest method of removing hydrogen sulphide is by water scrubbing of the biogas. This also removes carbon dioxide making the gas richer in methane (as does passage through an alkaline adsorbent), but there are no advantages in most plants. Passage through iron oxide sponge is a common method of chemically removing H_2S and this has been advocated for farm plants (Ravishanker and Hills, 1984): Kayhanian and Hills (1988) have also discussed membrane purification of biogas.

The biogas can be used at a few centimetres water gauge pressure in a boiler or in an engine, or, on the small developing-country farm, for cookers and gas lights. Boilers can be standard natural-gas equipment with the jets slightly enlarged. A boiler is usually arranged with two hot water circuits, one heats the digester as demanded by thermostats in the digester tank, the other heats animal or farmhouses. Such circuits can be arranged in conjunction with dual-fuel burner boilers or separate oil boilers which can take over the heating if the biogas supply is for any reason insufficient. Engines for farm plants are usually spark-ignition engines converted from petrol to gas running; they can be from about 15 kW electrical output upwards. Large sewage works generally use dual-fuel engines running on gas ignited by diesel oil with compression ignition, but such engines are too large for most farm plants. Although an engine can provide motive power for farm equipment the most usual use is electricity generation. The electricity can be used on the farm as well as to power pumps and other equipment on the digester. There are some problems finding sufficient uses for the electricity which will ensure that the generator is running constantly at its most efficient range (near its maximum power output). The generator also needs to cope with peak power demands when all the animal-house heating or ventilation is required or when milking or feed-milling or other activities take place. Operation such that mains electricity can take part of the peak load is often arranged. It is also becoming easier and economically more attractive to sell power back to the electricity authorities during periods of low load demand. This can help to even out electricity generation. The same comments about evening out the load apply to gas use in boilers.

3.8.2 Heating the digester

The most common heating systems are based on hot-water heat-exchangers in the digesting slurry, although some digesters pass the feedstock through a heat-exchanger before it enters the digester, or circulate the digester contents through an external heat-exchanger. Internal heat-exchangers can be of different types, there is no standard system. The hot water is supplied from a biogas boiler, which may have other uses, or from a gas engine. The gas engines have the usual cooling jackets for the cylinder block and also heat-exchange water jackets on the exhaust system. Some small engine systems have the entire engine and generator encased in an insulated container so that the maximum heat transfer can be obtained. Such combined heat and power units can have an overall fuel efficiency as electricity and hot water of about 90%. The engines may or may not have conventional air radiators to cope with periods of low heat demand elsewhere.

3.8.3 Mixing

There are three common systems of digester mixing; a mechanical impeller working free, or in a draft tube; gas-lift mixing, the gas similarly rising freely or confined in a draft tube; and pumped recirculation of digester contents. It is usual to position the mixer(s) and heat-exchanger(s) near to each other to maximize heat transfer and prevent the drying of solids on the heat-exchanger surfaces. This would otherwise reduce heat-exchange efficiency. Double-skinned draft-tube heat-exchangers and mixers are self-cleaning on the inside faces of the tubes.

3.8.4 Feeding

For below-ground digesters feeding can be by gravity flow. Outflow may also be by gravity and the rise in level caused by the inflow, or by pump on the larger plants. Above-ground digesters require a pumped feed, the pump being controlled by timer or level of the digester contents or both. Slurry pumps are usually scroll-in-flexible-stator types but large peristaltic pumps are also available.

Level sensors are usually fitted as an override to timers in case of failure of the 'off' cycle. Output can be by gravity overflow as the feed raises the level of the tank contents or by a pump working in time with the feed pump. However, in the case of a pumped outlet then a level detector is also needed as the feed and effluent streams cannot be relied on to exactly balance. Overfilling or the creation of a vacuum in the tank due to too great an outflow could occur. Over-pressure of the digester due to gas-line blockages or liquid overfill is usually prevented by a safety valve, often a simple liquid-filled U-tube in the digester roof. This will also relieve negative pressures but at the cost of letting air into the tank. Admission of some air through a leaking feed pipe or other cause can be tolerated as the bacteria have a large capacity to take up oxygen, but the gas then becomes diluted with nitrogen (Hobson *et al.*, 1974) and may become non-combustible.

3.9 Residues after digestion

The main end-use of farm digester slurry is as a fertilizer utilizing the nitrogen, phosphorus and potassium (NPK) values retained from the original slurry, but with pollutants (including odour), pathogenic organisms and weed seeds eliminated or considerably reduced. The reduction in biodegradable organic matter also improves the slurry as a fertilizer. Digested slurry is very stable and can be stored for long periods until the weather or crop cycle will allow land spreading. Solids can be separated from digested slurry for use as fertilizer and the liquid can be used for irrigation or might be treated further for discharge to a water course. The ammonia in digested slurry effluent is a particular problem for later aerobic treatment. Ammonia is a fertilizer on the land, but in natural waters it induces algal growth and is oxidized to nitrate. Both are undesirable. Forced aeration can strip off ammonia, or it can be removed by nitrifying–denitrifying cycles in aeration plants.

In warm climates the digester effluent has been used for algal production in ponds, the algae are harvested as a protein source for animal feeds or used in a sequence of ponds as fish food. The digested waste itself can also be used for fertilizing fish ponds. The separated digested solids contain a high proportion of bacterial cells which can form the basis of an animal-feed protein supplement. If such protein were to be made on a large scale in industrialized countries then the protein would have to be sterilized before distribution to comply with various requirements for bacterial content of animal feeds. The sterilization would not necessarily be required for feed processed and used on the same farm, but there could then be the risk of recycling disease within the farm. The production of animal feeds is bound up with the price and availability of conventional plant-, fish- and animal-feeds. The production of protein from digester effluent is usually uneconomic in view of the prices of the conventional feeds in many countries. The water from digester effluents could be recycled for wash water or animal drinking water within the farm. But here, again, there could be problems of disease and also of build-up of salts in the drinking water. These could upset the animals and the digester bacteria as the recycled water would eventually return to the digester and the recycling would be continuous. Some experiments by the author's group did show that liquid from a piggery digester was taken by fattening pigs without ill effects when partially diluted with fresh water. This was a short-term test and more extensive experiments would be needed before recommendations for recycling could be made. Chang and Fairbanks (1981) reviewed treatments of, and uses for, wastewaters from digester effluents, but the subject is mentioned in many other papers, including those on digester running.

3.10 Modelling

Two types of mathematical model of digesters can be made and both can have a place in optimizing farm digestion. One type of model is an economic model which can integrate costs and benefits of different digester constructions, pollution control and uses of gas and end-products to determine the optimum digestion plant for a particular farm or perhaps for a group of farms where waste treatment is centralized. The other type of model is one which seeks to describe the digestion process in terms of the biochemical reactions and the growth of the bacteria to try to optimize the digestion process itself and/or gain a better understanding of the microbial activities.

The simplest of the economic models is a preliminary calculation of the approximate size of the digester and its possible outputs. Proceeding from this the results of detailed surveys of the farm and the slurry-handling system give more exact figures for the volumes of wastes to be treated and the flow patterns to and from the digester. The following procedures are more or less standard civil and mechanical engineering site-practice. Ideally it may be possible to provide a standard digester coupled to standard engine-generator and slurry-handling 'packages'. Alternatively the plant may have to be designed on a 'one-off' basis. The uses of the gas and effluent have to be decided on as the digester is designed. The consideration of all these economic factors and the drawing of plans and costings is usually done by making numbers of calculations without any formal model structure. However a number of computer programs have also been proposed into which all the variables of treatment and construction can be entered and from which the various alternatives can be compared in economic and other terms. Whether any of these programs have been used commercially is not known.

The 'biological' model can be of two basic types. One describes the reactions and the results in terms of gas output and purification of the slurry when the digester is running at 'steady-state'. This defines the digester as running at a constant feed input, constant temperature and also giving a constant output of digested effluent and gas. The other type of model is a 'dynamic' one which gives an output showing the progress of digestion with time as the digester is started up, as it runs at steady-state, or what happens if temperature, loading rate or another parameter is changed. Just as the economic model has to have an input of costs of materials and other factors to give a comprehensive answer to questions posed, so the biological model has to have an input of amounts of feed, its degradability and other facts. Before the advent of the computer, time or impossibility of calculation was a major factor limiting the application of biological models. This is no longer the case, but the limiting factor is the availability of data on the rates and extents of reactions and the influences controlling bacterial growth in the digester. Biological models always present a more or less simplified picture of microbial reactions, nevertheless useful results for considerations of digester design and running have been obtained. The literature on modelling is too extensive to be considered here, but a short list of some relevant papers is given in the Bibliography.

3.11 Conclusions

Demands for the treatment of wastes from intensive farms and from industries associated with agriculture and food preparation are increasing. Anaerobic digestion offers a reliable method of reducing pollution from such wastes while at the same time providing a return on the costs of pollution control in the form of biogas. Even if this biogas can only provide the equivalent of the energy needed to run the plant then the pollution control provided by anaerobic digestion costs nothing to run. Whether gassification by anaerobic digestion of wastes or plant material grown specially for energy will become an important 'alternative energy' source is not known. At present the benefits of digestion are for pollution control.

Of less immediate benefit to the farmer, but of significance in the future is the energy efficiency of the system. In mechanical aerobic treatment carbonaceous pollutants are oxidized to carbon dioxide which escapes to the atmosphere. To do this conversion needs large inputs of electrical energy generally generated by burning fossil fuels to carbon dioxide. In anaerobic treatment some 50% of this pollutant carbon in the waste is converted to methane. The burning of this methane in an engine-generator does produce carbon dioxide, but the electricity can run the plant and obviate the need for external electricity generation. The heat also produced can replace the burning of oil for heating the animal- or farmhouses, which would be needed with an aerobic treatment plant for the farm wastes.

Present-day anaerobic digesters are not confined to treatment of slurries but can reduce pollution from all types of wastes, from solids to wastewaters. Digestion would seem to be a strong contender for present and future pollution-control plants for farms.

3.12 References

BABA (1982). *Anaerobic digesters. A code of practice on safety in and around anaerobic digesters.* BABA Ltd, Reading, UK.

Badger, D. M., Bogue, M. J. and Stewart, D. J. (1979). Biogas production from crops and organic wastes. *New Zealand Journal of Science*, **22**, 11–20.

Balsari, P. and Bozza, E. (1988). Fertilizers and biogas recovery installation of a slurry lagoon. In *Agricultural Waste Management and Environmental Protection*, ed. E. Welte and I. Szabolcs, CIEC-FAL, Braunschweig, West Germany, pp. 71–80.

Barry, M. and Colleran, E. (1982). Anaerobic digestion of silage effluent using an upflow fixed film reactor. *Agricultural Wastes*, **4**, 231–9.

van den Berg, L. and Lentz, C. P. (1979). Comparison between up- and down-flow anaerobic fixed-film reactors of varying surface to volume ratios in the treatment of bean-blanching waste. *Proceedings of the 34th Purdue Industrial Waste Conference*, pp. 319–25.

Brouard, F., Bories, A. and Stauze, F. (1983). Advance in anaerobic digestion of aquatic plants. In *Energy from Biomass*, ed. A. Strub, P. Chartier and G. Schleser. Applied Science Publishers, London, pp. 334–8.

Callander, I. J. and Barford, J. P. (1984). Improved anaerobic digestion of pig manure using a tower fermenter. *Agricultural Wastes*, **11**, 1–24.

Chang, A. C. and Fairbanks, W. C. (1981). A study on the treatment and disposal of waste water generated by methane producing anaerobic digesters. Report prepared for Pacific Gas and Electric Co., San Francisco, USA.

Chapman, J. M., Hawkes, D. L. and Pain, B. F. (1988). Performance studies of pilot-scale inclined tubular digesters operating on cattle slurry. In *Fifth International Symposium on Anaerobic Digestion: Poster Papers*, ed. A. Tilche and A. Rozzi. Monduzzi, Bologna, Italy, pp. 375–8.

Chen, Y. R. (1983). Kinetic analysis of anaerobic digestion of pig manure and its design implications. *Agricultural Wastes*, **8**, 65–81.

Clausen, E. C. and Gaddy, J. L. (1983). Methane production from agricultural residues by anaerobic digestion in batch and continuous culture. In *Fuel Gas Systems*, ed. D. Wise, CRC Press, Boca Raton, Florida, USA, pp. 111–40.

Coble, C. G. and Egg, R. (1987). Ensilage storage of sorghum and high moisture biomass crops for anaerobic digestion feedstocks. In *Energy from Biomass and Wastes X*, ed. D. L. Klass, Elsevier Applied Science Publishers and Institute of Gas Technology, London and Chicago, pp. 1057–76.

Cohen, A., Zoetemeyer, R. J., van Deursen, A. and Andel, J. G. (1979). Anaerobic digestion of glucose with separated acid and methane formation. *Water Research*, **13**, 571–80.

Colleran, E., Barry, M., Wilkie, A. and Newell, P. J. (1982). Anaerobic digestion of agricultural wastes using the upflow anaerobic filter design. *Process Biochemistry*, **17**(2), 12–17.

Cullimore, D. R., Maule, A. and Mansuy, N. (1985). Ambient temperature methanogenesis from pig manure waste lagoons. *Agricultural Wastes*, **12**, 147–57.

Dar, G. H. Hassan and Tandon, S. M. (1987). Biogas production from pretreated wheat straw, Lantana residue, apple and peach leaf litter with cattle dung. *Biological Wastes*, **21**, 75–84.

Demuynck, M., Nyns, E. J. and Paltz, W. (1984). *Biogas Plants in Europe. A practical handbook*. D. Reidel Publishing Company, Dordrecht.

Feilden, N. E. H. (1981). A note on the temperature for maximum net gas production in an anaerobic digester system. *Agricultural Wastes*, **3**, 75–9.

Field, J. A., Reneau, R. B., Kroontje, W. and Caldwell, J. S. (1985). Nutrient recoveries from plug-flow anaerobic digestion of poultry manure. *Agricultural Wastes*, **13**, 207–16.

Floyd, J. R. S. and Hawkes, F. R. (1986). Operation of a laboratory-scale tubular digester on piggery waste. *Agricultural Wastes*, **18**, 39–60.

Fry, L. J. (1974). *Practical Building of Methane Power Plants*. A. Knox, Andover, UK.

Ghosh, S. (1984). Advanced two-phase digestion of sewage sludge. In *Energy from Biomass and Wastes VIII*, Institute of Gas Technology, Chicago, USA, pp. 853–74.

Guo-Chao Sun, Yi-Zhe Wu, Shi-Jin Sha and Ke-Xin Liu (1987). Dry digestion of crop wastes: studies on dry anaerobic digestion with agricultural wastes. *Biological Wastes*, **20**, 291–302.

Haga, K., Tanaka, H. S. and Higaki, S. (1979). Methane production from animal wastes and its prospects in Japan. *Agricultural Wastes*, **1**, 45–56.

Hall, S. J., Hawkes, D. L., Hawkes, F. R. and Thomas, A. (1985). Mesophilic anaerobic digestion of high-solids cattle waste in a packed-bed digester. *Journal of Agricultural Engineering Research*, **32**, 153–62.

Hashimoto, A. G. (1983). Thermophilic and mesophilic anaerobic fermentation of swine manure. *Agricultural Wastes*, **6**, 175–91.

Hayes, T. D., Jewell, W. J., Dell'Orto, S., Fanfoni, K. J., Leuschner, A. P. and Sherman, D. F. (1980). Anaerobic digestion of cattle manure. In *Anaerobic Digestion*, ed. D. A. Stafford, B. I. Wheatley and D. E. Hughes. Applied Science Publishers, London, pp. 255–88.

Hilbert, R., Winter, J. and Kandler, O. (1984). Agricultural feed additives and disinfectants as inhibitory factors in anaerobic digestion. *Agricultural Wastes*, **10**, 81–94.

Hill, D. T. and Bolte, J. P. (1986). Characteristics of whole and scraped swine waste as substrates for continuously expanding anaerobic digestion systems. *Agricultural Wastes*, **16**, 147–56.

Hill, D. T., Young, D. T. and Norstedt, R. A. (1981). Continuously expanding anaerobic digestion—a technology for the small animal producer. *Transactions ASAE*, **24**, 731–6.

Hill, D. T., Holmeberg, R. D. and Bolte, J. P. (1985). Operating and performance characteristics of scraped swine manure as a thermophilic anaerobic digestion substrate. *Agricultural Wastes*, **14**, 37–50.

Hills, D. J. (1979). Effects of carbon: nitrogen ratio on anaerobic digestion of dairy manure. *Agricultural Wastes*, **1**, 267–78.

Hobson, P. N. (1979). Straw as a feedstock for anaerobic digesters. In *Straw Decay and its Effect on Disposal and Utilisation*, ed. E. Grossbard. John Wiley, Chichester, pp. 217–26.

Hobson, P. N. (1984). Mathematical models. In *Models of Anaerobic Infection*, ed. M. J. Hill. Martinus Nijhoff, Dordrecht, pp. 151–64.

Hobson, P. N. and Feilden, N. E. H. (1982). Production and use of biogas in agriculture. *Prog. Energy Combustion Science*, **8**, 135–58.

Hobson, P. N. and Shaw, B. G. (1973). The anaerobic digestion of waste from an intensive pig unit. *Water Research*, **7**, 437–49.

Hobson, P. N., Bousfield, S. and Summers, R. (1974). The anaerobic digestion of organic matter. *Crit. Rev. Environ. Cont.*, **4**, 131–91.

Hobson, P. N. and Robertson, A. M. (1977). *Waste Treatment in Agriculture*. Applied Science Publishers, London.

Hobson, P. N., Bousfield, S. and Summers, R. (1981). *Methane Production from Agricultural and Domestic Wastes*. Applied Science Publishers, London.

Hobson, P. N., Bousfield, S., Summers, R. and Mills, P. J. (1980). Anaerobic digestion of piggery and poultry wastes. In *Anaerobic Digestion*, ed. D. A. Stafford, B. I. Wheatley and D. E. Hughes. Applied Science Publishers, London, pp. 237–54.

Hobson, P. N., Summers, R. and Harries, C. (1984). Single- and multi-stage fermenters for treatment of agricultural wastes. In *Microbiological Methods for Environmental Biotechnology*, ed. J. M. Grainger and J. M. Lynch. Academic Press, London, pp. 119–38.

Horton, R. (1980). The implications of engineering design on anaerobic digester systems. In *Anaerobic Digestion*, ed. D. A. Stafford, B. I. Wheatley and D. E. Hughes. Applied Science Publishers, London, pp. 321–44.

Jewell, W. J. (1980). Future trends in digester design. In *Anaerobic Digestion*, ed. D. A. Stafford, B. I. Wheatley and D. E. Hughes. Applied Science Publishers, London, pp. 467–92.

Kayhanian, M. and Mills, D. J. (1988). Membrane purification of anaerobic digester gas. *Biological Wastes*, **23**, 1–16.

Kennedy, K. J. and van den Berg, L. (1982). Anaerobic digestion of piggery waste using a stationary fixed film reactor. *Agricultural Wastes*, **4**, 151–8.

Lettinga, G., van Velsen, A. F. M., de Zeeuw, W. and Hobma, S. W. (1979). Feasibility of the upflow anaerobic sludge blanket (UASB) process. *Proceedings, National Conference on Environmental Engineering*, ASCE, New York, pp. 35–42.

Lo, K. V., Bulley, N. R., Liao, P. H. and Whitehead, A. J. (1983). The effect of solids-separation pretreatment on biogas production from dairy manure. *Agricultural Wastes*, **8**, 155–66.

Loehr, R. C. (1984). *Pollution Control for Agriculture*. Academic Press, New York.

McKain, N. and Hobson, P. N. (1987). A note on the destruction of porcine enteroviruses in anaerobic digestions. *Biological Wastes*, **22**, 147–55.
Molnar, L. and Bartha, I. (1989). Factors influencing solid-state anaerobic digestion. *Biological Wastes*, **28**, 15–24.
Morrison, I. M. (1979). Degradation and utilisation of straw in the rumen. In *Straw Decay and its Effect on Disposal and Utilisation*, ed. E. Grossbard. John Wiley & Sons, Chichester, pp. 237–46.
Noone, G. P. and Boyd, A. K. (1980). Prefabricated systems for low-cost anaerobic digestion. EEC Symposium on sludge digestion. COST bis 68. Vienna.
Oleszkiewicz, J. A. (1983). A comparison of anaerobic treatments of low concentration piggery wastewaters. *Agricultural Wastes*, **8**, 215–32.
Oleszkiewicz, J. A. (1985). Cost-effective treatment of piggery wastewater. *Agricultural Wastes*, **12**, 185–206.
Oleszkiewicz, J. A. and Kosiarski, S. (1986). Kinetics of piggery wastes treatment in anaerobic lagoons. *Agricultural Wastes*, **16**, 13–26.
Olsen, J. E., Jorgensen, J. B. and Nansen, P. (1985). On the reduction of *Mycobacterium paratuberculosis* in bovine slurry subjected to batch mesophilic or thermophilic anaerobic digestion. *Agricultural Wastes*, **13**, 273–80.
Pain, B. F., West, R., Oliver, B. and Hawkes, D. L. (1984). Mesophilic anaerobic digestion of dairy cow slurry on a farm scale. First comparisons between digestion before and after solids separation. *Journal of Agricultural Engineering Research*, **29**, 249–56.
Palz, W. and Cartier, P. (1980). *Energy from Biomass in Europe*. Applied Science Publishers, London.
Plym-Forshell, L. (1983). Survival of salmonella bacteria and *Ascaris suum* eggs in a thermophilic biogas plant. In *Hygienic Problems of Animal Manures*, ed. D. Strauch. Institute of Animal Medicine and Hygiene, University of Hohenheim, Stuttgart, Germany, pp. 217–22.
Poels, J., Neukermans, G., van Assche, P., Debruyckere, M. and Verstraete, W. (1983). Performance, operation and benefits of an anaerobic digestion system on a closed piggery farm. *Agricultural Wastes*, **8**, 233–49.
Poels, J., van Assche, P. and Verstraete, W. (1984). Effects of disinfectants and antibiotics on the anaerobic digestion of piggery waste. *Agricultural Wastes*, **9**, 239–48.
Ravishanker, P. and Hills, D. J. (1984). Hydrogen sulphide removal from anaerobic digester gas. *Agricultural Wastes*, **11**, 167–80.
Robbins, J. E., Arnold, M. J. and Weiel, J. E. (1983). Anaerobic digestion of cellulose-dairy cattle manure mixtures. *Agricultural Wastes*, **8**, 105–18.
Robbins, J. E., Gerhardt, A. and Kappel, T. J. (1989). Effects of total ammonia on anaerobic digestion and an example of digester performance from cattle manure-protein mixtures. *Biological Wastes*, **27**, 1–14.
Safley, L. M., Vetter, R. L. and Smith, D. (1987). Operating a full-scale poultry manure anaerobic digester. *Biological Wastes*, **19**, 79–90.
Safley, L. M. and Westerman, P. W. (1989). Anaerobic lagoon biogas recovery system. *Biological Wastes*, **27**, 43–62.

Sanders, D. A., Malina, J. F. Jr, Moore, B. E., Sagik, B. P. and Sorber, C. A. (1979). Fate of polio virus during anaerobic digestion. *Journal of the Water Pollution Control Federation*, **51**, 333–43.

Sharma, Sudhir K., Saini, J. S., Mishra, I. M. and Sharma, M. P. (1989). Biogassification of woody biomass: *Ipomoea fistulosa* plant stem. *Biological Wastes*, **28**, 25–32.

Stanogias, G., Tjandraatmadja, M. and Pearce, G. R. (1985). Effects of source and level of fibre in pig diets on methane production from pig faeces. *Agricultural Wastes*, **12**, 37–54.

Summers, R. and Bousfield, S. (1980). A detailed study of piggery waste anaerobic digestion. *Agricultural Wastes*, **2**, 61–78.

Summers, R., Hobson, P. N., Harries, C. and Feilden, N. E. H. (1984). Anaerobic digestion on a large pig unit. *Process Biochemistry*, **19**, 77–8.

Summers, R., Hobson, P. N., Harries, C. and Richardson, A. J. (1987). Stirred-tank, mesophilic, anaerobic digestion of fattening cattle wastes and of whole and separated dairy-cattle wastes. *Biological Wastes*, **20**, 43–62.

Sutter, K. and Wellinger, A. (1988). The ACF system; a new low-temperature biogas digester. In *Agricultural Wastes Management and Environmental Protection*, ed. E. Welte and I. Szabolcs, CIEC-FAL, Braunschweig, W. Germany.

Turner, J., Stafford, D. A., Hughes, D. E. and Clarkson, J. (1983). The reduction of three plant pathogens (*Fusarium*, *Corynebacterium* and Globodera) in anaerobic digesters. *Agricultural Wastes*, **6**, 1–12.

Varel, V. H. and Hashimoto, A. G. (1981). Effect of dietary monensin and chlortetracycline on methane production from animal wastes. *Applied and Environmental Biotechnology*, **41**, 29–34.

Varel, V. H., Hashimoto, A. G. and Chen, Y. R. (1980). Effect of temperature and retention time on methane production from cattle waste. *Applied and Environmental Microbiology*, **40**, 217–22.

Varel, V. H., Isaacson, H. R. and Bryant, M. P. (1977). Thermophilic methane production from cattle waste. *Applied and Environmental Microbiology*, **33**, 298–307.

van Velsen, A. F. M. (1977). Anaerobic digestion of piggery waste. 1. The influence of detention time and manure concentration. *Netherlands Journal of Agricultural Science*, **25**, 151–69.

van Velsen, A. F. M. (1979). Adaptation of methanogenic sludge to high ammonia-N concentrations. *Water Research*, **13**, 995–9.

van Velsen, A. F. M., Lettinga, G. A. and Ottelander, D. (1979). Anaerobic digestion of piggery waste 3. Influence of temperature. *Netherlands Journal of Agricultural Science*, **27**, 255–67.

Viturtia, A. M., Mata-Alvarez, J., Cecchi, F. and Fazzini, G. (1989). Two-phase anaerobic digestion of a mixture of fruit and vegetable wastes. *Biological Wastes*, **29**, 189–200.

Wellinger, A. and Kaufmann, R. (1982). Psychrophilic methane generation from pig manure. *Process Biochemistry*, **17**(5), 26–30.

Wong, M. H. and Cheung, Y. H. (1989). Anaerobic digestion of pig manure with different agro-industrial wastes. *Biological Wastes*, **28**, 143–56.

Yang, P. Y. (1981). Methane fermentation of Hawaiian seaweeds. In *Energy from Biomass and Wastes V*, Institute of Gas Technology, Chicago, USA, pp. 307–27.
Young, J. C. and McCarty, P. L. (1969). The anaerobic filter for waste treatment. *Journal of the Water Pollution Control Federation*, **41**(15), R160–73.
Zeeman, G., Wiegant, W. M., Koster-Treffers, M. E. and Lettinga, G. (1985). The influence of total ammonia concentration on the thermophilic digestion of cow manure. *Agricultural Wastes*, **14**, 19–36.
Zeeman, G., Sutter, K., Vens, T., Koster, M. and Wellinger, A. (1988). Psychrophilic digestion of dairy cattle and pig manure: start-up procedures for batch, fed-batch and CSTR-type digesters. *Biological Wastes*, **26**, 15–32.

3.13 Bibliography

Papers on modelling and digester function

Chen, Y. R. (1983). Kinetic analysis of anaerobic digestion of pig manure and its design implications. *Agricultural Wastes*, **8**, 65–81.
Chen, Y. R. and Hashimoto, A. G. (1978). Kinetics of methane fermentation. *Biotechnology and Bioengineering Symposium No. 8*, 269–82.
Hanaki, K., Noike, T. and Matsumoto, J. (1985). Mathematical modelling of the anaerobic digestion process. *Developments in Environmental Modelling*, **7**, 583–636.
Hill, D. T. (1982). A comprehensive dynamic model for animal waste methanogenesis. *Transactions of the ASAE*, **25**, 1374–80.
Hill, D. T. and Norstedt, R. A. (1980). Mathematical modelling and computer simulation of agricultural waste treatment processes. *Agricultural Wastes*, **2**, 135–56.
Hobson, P. N. (1983). The kinetics of anaerobic digestion of farm wastes. *Journal of Chemical Technology and Biotechnology*, **33b**, 1–20.
Hobson, P. N. (1985). A model of anaerobic bacterial degradation of solid substrates in a batch digester. *Agricultural Wastes*, **14**, 255–74.
Hobson, P. N. (1987). A model of some aspects of the microbial degradation of particulate substrates. *Journal of Fermentation Technology*, **65**, 431–9.
Hobson, P. N. and McDonald, I. (1980). Methane production from acids in piggery waste digesters. *Journal of Chemical Technology and Biotechnology*, **30**, 405–8.
Lawrence, A. W. and McCarty, P. L. (1969). Kinetics of methane fermentation in anaerobic treatment. *Journal of the Water Pollution Control Federation*, **41**, R1–R7.
Lo, K. V., Carson, W. M. and Jeffers, K. (1981). A computer aided design program for biogas production from animal manure. In *Livestock Wastes: a Renewable Resource*. St. Joseph, USA, American Society of Agricultural Engineers.

Mosey, F. E. and Fernandez, X. A. (1984). Mathematical modelling of methanogens in sewage sludge digestion. In *Microbiological Methods for Environmental Biotechnology*, ed. J. M. Grainger and J. M. Lynch. Academic Press, London, pp. 159–68.

Rozzi, A. and Passino, R. (1985). Mathematical models in anaerobic treatment processes. *Developments in Environ. Modelling*, **7**, 637–90.

Books on digesters, and the digestion of agro-food wastes

van Buren, A. (ed.) (1979). *A Chinese Biogas Manual*. Intermediate Technology Publications Ltd, London (Chinese digesters).

Chynoweth, D. P. and Isaacson, R. (eds) (1987). *Anaerobic Digestion of Biomass*. Elsevier Applied Science Publishers, London.

Energy from Biomass and Wastes (volumes 1 onwards 1976 and continuing). Proceedings of symposia sponsored by The Institute of Gas Technology, Chicago, USA.

Ferrero, G. L., Ferranti, M. P. and Naveau, H. (eds) (1984). *Anaerobic Digestion and Carbohydrate Hydrolysis of Waste*. Elsevier Applied Science, London.

Meynell, P.-J. (1982). *Methane—Planning a Digester*, 2nd edition. Prism Press, Dorchester.

Pain, B. F. and Hepherd, R. Q. (eds) (1985). *Anaerobic Digestion of Farm Waste*. National Institute for Research in Dairying, Reading, UK.

4 The treatment of domestic wastes

G. P. Noone
Severn–Trent Water Authority, Birmingham, UK

4.1	**Introduction**	140
4.2	**The history of anaerobic digestion**	143
4.3	**Process design**	145
4.3.1	Background	145
4.4	**Process equipment**	146
4.4.1	Mixing	146
	(a) Sludge recirculation	147
	(b) Gas recirculation	148
	(c) Impeller mixers	148
	(d) Power requirements and efficiency	148
4.4.2	Heating	150
	(a) External heat exchangers	151
	(b) Heat transfer and efficiency	151
	(c) Steam injection	152
	(d) Direct combustion	152
4.4.3	Sludge thickening	153
	(a) Sludge consolidation	153
	(b) Sludge consolidation within primary tanks	155
4.4.4	Sludge feeding	155
4.4.5	Gas storage	155
4.5	**The digester**	156
4.5.1	Retention time	157
4.5.2	Prefabrication	159
4.5.3	Aspect ratio	161
4.6	**The residues after digestion**	161
4.7	**The routine operation**	162
4.7.1	Control and monitoring	163
4.7.2	Analysis	163
	(a) Reduction of organic matter	163
	(b) pH, alkalinity and carboxylic acids	164

4.7.3   Effective biomass 165
4.7.4   Inhibitors 165

4.8   Design and procurement procedures 166

4.9   Future trends 167

4.10   Conclusions 168

4.11   Acknowledgements 168

4.12   References 168

4.1   Introduction

Solids liquid separation and sludge production is an integral part of the domestic sewage treatment processes (Fig. 4.1). The surplus solids or sludges produced by settlement are putrescible and offensive. Treatment is essential to avoid causing a nuisance during disposal.

In all countries practising complete sewage treatment, mesophilic (30–35°C) anaerobic digestion is the most popular type of treatment for these sludges. In the UK, 60% of sludge is anaerobically digested (Table 4.1 taken from Noone, 1984); this is comparable to other European countries and similar to the average in North America (Vincent and Critchley, 1984).

Anaerobic treatment has also been considered and tested for other stages of the sewage treatment process (Hall and Hobson, 1988 and Chapter 1), particularly in warm countries where the ideal mesophilic temperatures required for successful anaerobic digestion may be

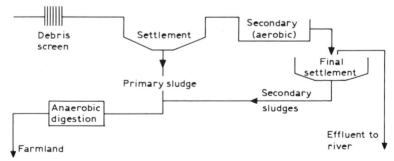

Fig. 4.1   Typical sewage/sludge treatment/sludge disposal flowsheet.

Table 4.1 Sludge stabilisation treatment in Europe

Country	Current raw sludge production (tonnes dry solids/year)	Use of stabilisation (% of sludge produced)					% stabilised before disposal	% sterilised before disposal
		None	Anaerobic	Aerobic	Compost	Lime		
UK	1 500	40	58	1	—	—	60	1
Belgium	70	19		81			81	1
Denmark	130	4	50	40	1	5	96	29
France	840	41	49	7	3		59	3
Germany	2 200	25	65	7	3	1	75	17
Greece	3	—	—	—	—	—	—	—
Ireland	20	62		38			38	1
Italy	1 200	25		75			75	
Luxembourg	11	—	—	—	—	—	Substantial	—
Netherlands	230	35	43	21	—	—	65	20
Austria	140	30		50	20		70	29
Finland	130	58	20	13	1	8	42	41
Norway	55	76	10	5	4	5	24	
Spain	45	—	—	—	—	—	—	—
Sweden	210	10	50	18	7	15	90	26
Switzerland	150	19	77	2	2	—	81	7
Europe	6 934	31		66	2	1	69	9
Canada	500	—	—	—	—	—	—	11

achieved without heating. These applications are still at a pilot or research stage and the most common full-scale use of anaerobic processes for domestic wastes is for the treatment of low volume/high strength wastes such as sludge.

The history and development of the anaerobic digestion process are detailed in this chapter. Digestion was developed originally because of its ability to control and eliminate the malodour associated with raw sludges. This remains the main reason for its widespread use.

Two types of sludge are produced during the treatment of domestic sewage (Fig. 4.1). Large and easily separable solids in sewage are removed early in the process by settlement tanks. This can remove 30% of the overall polluting load of sewage as well as reduce the possibility of blockage and clogging later in the treatment process. The clarified surface liquid then proceeds to a secondary (invariably aerobic) treatment stage. This aerobic biological treatment of the liquid phase generates surplus biomass which is separated by a second settlement tank. These are secondary sludges. It is normal to combine primary and secondary sludge together with any additional sludges from further or polishing treatment stages. The combined works sludges are then subjected to a single sludge treatment and subsequent disposal route.

Proposed new guidelines and legislation on the disposal and use of sewage sludge (CEC, 1981) has increased the interest in the application of anaerobic digestion. Much of the sludge not anaerobically treated is disposed of directly to land or sea (Table 4.1). Restrictions on sea disposal have increased significantly. In addition to odour control, anaerobic digestion has other beneficial effects:

(1) Reduction in the number of pathogens.
(2) Reduction in solids content.
(3) Methane generation as a by-product.
(4) Improved solubilisation of nitrogen for fertilising action.

The need for digestion to give disposal flexibility figures prominently in the guidelines for sewage sludge disposal. The warm temperatures (35°C—mesophilic range) and the anaerobic conditions are key factors in the reduction of pathogens. Thus anaerobically treated sludge is free from odour and more hygienic. Each of these make further sludge processing or disposal to land easier.

4.2 The history of anaerobic digestion

The treatment of sewage sludges was the first application of anaerobic processing. Full-scale anaerobic digestion was practised long before the microbial and biochemical understanding had developed sufficiently for effective control. Methane production from sewage sludge was noted in 1885 by Dibdin. The collection and harnessing of this useful by-product gas for works lighting was reported by Cameron at Exeter in 1895 (Stanbridge, 1976).

The early 'processes' prior to 1910 were without heating or mixing and were thus merely variations upon sludge storage systems.

It is not clear from the literature whether the first digesters with heating and mixing were built in Germany or the UK. Whitehead and O'Shaughnessy (1931) reported on the beneficial effects of heating and mixing in some experiments they had been conducting on sludge digestion in Birmingham. Whitehead and O'Shaughnessy particularly noted that it was much more difficult to get good mixing and, therefore, digestion at a large scale. Disposal of sewage sludges to land and sea was then relatively easy and unrestricted at this time and so little research or development was carried out on the process. In 1954 a Ministry of Local Government Working Party reported that the design of anaerobic digesters was very variable and 'unscientific'. It was however noted that dilute sludges caused problems by reducing retention time and gas production; also it was noted that good mixing was critical. Attempts were then made to improve the process during five years of investigation by the then Water Pollution Research Laboratory WPRL (Department of Scientific and Industrial Research, 1955–1960).

Optimum specific loading parameters were described in terms of volatile (reactable) solids per unit volume and per unit time. Research was carried out to determine the correct retention time (solids retention time) to prevent washout of the much slower growing methane-generating bacteria. The WPRL in a later part of this programme (Swanwick et al., 1969), conducted a National Survey of Digestion. The report indicated that inhibition of the 'fastidious' anaerobic microorganisms by toxins such as detergents, metals and solvents was not as common as believed. Many of the difficulties which had been noted were actually attributable to process design defects particularly inadequate heating and mixing. Contrary to prior expectation the survey showed that process difficulties were most often associated with long detention (>30 days) and lowly loaded digesters whilst those of 'medium' loading (<25 days) only experienced 'intermittent' difficulties whilst short detention (<15 days), highly loaded plants had few, if any, operational difficulties. This observation that higher reaction intensities lead to a more stable operation was controversial but sadly unnoticed because of the widely held misconception that 'bigger', long detention time reactors were safer and more reliable.

By 1975 the cost of sludge treatment and disposal was an acute problem and attention once again focussed on possible methods for improving anaerobic digestion. This led to the programme of extensive development work within the Severn–Trent Water Authority (Brade et al., 1981, 1982, 1984). The need for some disinfection of sewage sludges before land spreading was recognised in the Authority policy and anaerobic digestion was thought to have some economic and environmental advantages. This was confirmed by the first paper (Brade and Noone, 1981) which showed that all the sludge treatment alternatives to digestion suffered similar or worse problems.

The starting point for further development of the process was a review of existing digesters, followed by full-scale developments to improve the reliability and efficiency of the associated process equipment.

Table 4.2 The history of design provision for anaerobic digesters

Year	Reference	Design parameter
1925	Baltimore, USA	1·05 cu. ft/person
1928	USA (Imhoff type)	1·4–2·4 cu.ft/person
1928	USA (Dorr type)	1–3·7 cu. ft/person
1928	Germany	0·75–2·3 cu. ft/person
1928	Bath, UK	3·1 cu. ft/person
1931	UK (generally)	3·1 cu. ft/person
1928	New South Wales	3·0 cu. ft/person
1954	UK Informal Working Party	1–2 ft^3/head with 28–30 days' detention for mixed, mesophilic
Early Practice USA		Total solids 1·75 lb/ft^3/month (0·93 kg m^3/day)
1969	UK WPRL (Range of Survey)	16 days' detention
1975	CIRIA	Minimum, detention of 12 days

The treatment of domestic wastes 145

The history of typical design parameters is summarised in Table 4.2. In some of these early reactors some competing objectives in the same tank were observed e.g. decanting surface liquor whilst mixing and hoping to achieve settlement.

Despite some doubts about the absolute comparability of the data in Table 4.2, there is however a change in emphasis from volume-based designs to loading and substrate throughput. Current design practice within the UK is based on both loading rate (kg solids m^3/day) and a minimum detention time necessary to avoid wash out (i.e. 10 days or more). In practice it is extremely difficult to induce overloading because this would require very thick sludges (>20% dissolved solids (DS)); as this solids concentration cannot be achieved with any current thickening and mixing technology, then detention time remains the more relevant design parameter for sewage sludge digestion.

4.3 Process design

4.3.1 Background

A review of the literature on anaerobic sludge digestion was undertaken to compile a Severn–Trent Water Authority anaerobic digestion design manual (Brade and Noone, 1981). The review revealed wide variations in the key reactor dimensions, and little if any standardisation had occurred since the WPRL Survey in 1969 (Swanwick et al., 1969). In Germany the practice was to have retention times between 20–30 days because this improved the dewaterability and stabilisation of the sludge (Imhoff, 1984). In the United States 20 days were typical (Farrell, 1984), whilst in the United Kingdom, 12–15 days were advocated (Swanwick et al., 1969). A detailed survey of Severn–Trent digesters was carried out. The survey found a 5-fold variation in retention time. The other important features to emerge from the investigations were the large variations in:

(1) mixer performance,
(2) the efficiency and type of heating used,
(3) the method of feeding and discharge.

The process engineering design should be based on a completely mixed continuously heated reactor. In practice this was never achieved. None of the plants surveyed had continuous mixing, and only one plant had the capability for continuous heating. The sludge feeding regimes also varied from a single feed of 20 min in a 24 h period to a maximum frequency of five feeds per 24 h but all in a 12 h period. The basic reactor process engineering requirements had been subjugated by the civil structural consideration of the 'vessel'. Reactor shape was a good example. Most of the reactors were shallow, large diameter and therefore difficult to mix tanks, better suited to sludge storage rather than use as completely mixed reactors. Most of the digesters in Europe were however tall tanks or eggshaped (Imhoff, 1984).

In the light of this background variability, especially in the fundamental area of reactor mixing, full-scale plant development work was required to produce an optimum process design. The Severn–Trent Water Authority had a large number of anaerobic digesters (approximately 53% of all sludges were then digested; currently 85% of sludge is digested). The Authorities then calculated capital expenditure shortfall for additional plant, based upon historical costings, was £25 million at 1980 prices. This investment backlog in digestion facilities was however significantly lower than most, if not all, of the other UK Water Authorities.

Another role of the development work was, therefore, to look carefully at the possibility of reducing capital costs as well as improving anaerobic digestion.

4.4 Process equipment

4.4.1 *Mixing*

Mixing of the reactor is required:

(1) To disperse the influent raw sludge within the actively digesting sludge, thereby promoting contact between substrate and microorganism.
(2) To maintain dispersion and effective use of the volume of the reactor, thereby eliminating any isolated pockets or dead reactor zones. This reduces the formation of scum and grit.
(3) To eliminate thermal stratification and maintain a uniform temperature throughout the reactor.

The degree of mixing required to satisfy these requirements is itself a combination of a number of contributing factors namely:

(a) External applied mixing power (dependent on type of process mixing equipment).
(b) 'Internal' mixing power developed by gas evolution from the actively digesting sludge.

There are three basic methods of mixing sludge in digesters (Fig. 4.2):

(1) recirculation of sludge
(2) recirculation of gas through the sludge
(3) 'internal' mechanical mixing of sludge

(a) *Sludge recirculation.* Mixing by sludge recirculation with either

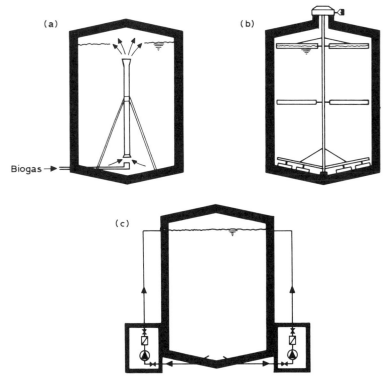

Fig. 4.2 The treatment of domestic waste. (a) Confined gas lift mixing; (b) impeller mechanical mixing; (c) recirculation of sludge.

conventional/mechanical pumping or by external gas lift pumps appears the least effective of the techniques available. This type of mixing circulates slugs of the digester contents rather than thoroughly mixing the sludge. Short circuiting and dead zones are therefore common problems with pumped recirculation. Sludge recirculation is now no longer used as the sole method of digester mixing.

(b) *Gas recirculation.* Gas mixers can be divided into two types: confined and unconfined. Gas is collected in the digester above the sludge surface, and is filtered and dewatered through a condensation trap, compressed and recycled back into the base of the digester. In the confined gas lift mixer, gas is released into a vertical tube below the sludge surface. The rising gas in the tube acts as a gas lift pump and sucks fresh sludge in at the base of the tube.

In unconfined gas lift mixing, the pressurised gas is simply released at say five or six evenly-spaced points around the base of the tank. Mixing occurs as the bubbles released rise to the surface of the tank.

(c) *Impeller mixers.* Internal mechanical mixers are of three types: roof mounted high speed, roof mounted low speed, and wall mounted high speed. The roof mounted high speed types were once the most common. The sludge was sucked up a fixed tube by a impeller and sprayed out over the digester surface. This action also helps to avoid scum problems. Occasionally the direction of the motor is reversed to prevent blockage of the tube. Power use in high speed mixers was high and comparable to compressor powers but less effective in operation.

(d) *Power requirements and efficiency.* There is little published information on the degree of self-mixing by sludge gassing. The application of self-mixing gas systems is confined to long residence time farm or rural digesters (see Chapters 1 and 3) although they are not effective. The external mixing power required for various typical sewage sludge digesters has been quantified (Rundle and Whyley, 1981; Noone and Brade, 1982). Principally, mixing efficiency depends on reactor aspect ratio and solids concentration. A summary of the results is shown in Fig. 4.3. The experimental derivation of the large amounts of data necessary to define Fig. 4.3 and thus totally optimise external power input is not justifiable.

Mixing design will remain empirically based and contain a degree of conservatism. It is usual to provide a larger than necessary amount of

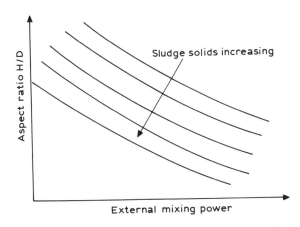

Fig. 4.3 Notional relationship of aspect ratio versus solids level versus external mixing power requirements.

external power which is then capable of downrating after process start up and some operating experience has been gained.

Some of the Severn–Trent digesters surveyed, for example, had both very poor and low external mixing power available. Plants with high solids concentrations and occasionally tall narrow shapes overcame this difficulty as adequate mixing was achieved by internal gassing.

Experimentally, the degree of mixing was examined by two parameters, 'digester turnover' which was the time taken to circulate the digester volume and 'digester dispersion' which was the time taken for an added tracer to reach an equilibrium concentration within the digester. The Severn–Trent experiments showed that unconfined gas recirculation gave the best dispersion (the relevant process mixing parameter) and least scum (Rundle and Whyley, 1981). Similar results have been presented in the USA (Baumann and Huibregtse, 1981) but results from Canada (Carroll and Ross, 1984) indicated superior dispersion from a confined gas lift system. Unconfined gas lift systems are however significantly cheaper in capital costs (Rundle and Whyley, 1981). The data on slow speed mechanical mixers showed them to be up to four times more efficient in terms of power requirements (Brade and Noone, 1981). Power requirements were found to be proportional to mixer peripheral velocity. Significant problems were however encountered with these mixers due to grit and rag accumulation from poor sludge screening. This detritus rapidly reduced the efficiency of mixing and then caused further accumulations of grit. Problems had also been encountered with the bottom bearings. Cleaning the impellers is expensive and unpleasant and this type of mixing is no longer thought of as suitable (Rundle and Whyley, 1981; Carroll and Ross, 1984).

Low speed dynamically balanced stirrers are also now available which eliminate the need for the bottom bearings. Early results show very good mixing efficiency (Noone and Brade, 1982), but that the mixing mechanism still collects rags and other debris. Satisfactory modern sludge screening has recently made this a reliable and effective mixing option. The current recommendation is for gas mixing as this has the advantage of no moving equipment within the tank. The minimum power provided is typically $6 \cdot 5 \times 10^{-3}\,\mathrm{Nm^3/m^2}$ (gas compressor continuous duty) at 4% solids, 15 days' retention, with a digester aspect ratio of 1:1.

4.4.2 Heating

It is well established that anaerobic microorganisms have an optimum growth temperature in the mesophilic 30–37°C and the thermophilic range 50–65°C (see Chapters 1 and 2). There are as yet few results from full-scale operating thermophilic plants although in theory the process should proceed more quickly at the higher temperature. Problems have however been encountered with process stability, high heating demands and odour release (Chapters 1 and 3). It seems unlikely that thermophilic digestion will be widely used except possibly for hot industrial wastes which would require no additional heating.

The heat transfer performance of any heat exchanger deteriorates with time due to fouling and scaling and the main process requirement for heating anaerobic digesters is for reliable performance. This is an important influence on the equipment chosen. Heat exchangers can be fitted both internally as coils within the digester or externally with pumped circulation. Internal heating coils have to be directly linked to the mixing system in the digester to avoid local overheating. These take the form of a jacketed gas lift tube or a coil around the impeller of the mixer. External heat exchangers are simpler to maintain, are more versatile and allow changes in heat transfer capacity but are much more expensive (Brade et al., 1982). In general external systems are to be preferred because of the ease of maintenance but the power and capital demands are significant. Steam heating offers the 'external' benefits but is used internally.

(a) *External heat exchangers.* There are two basic types:

CONCENTRIC TUBE EXCHANGER. This is the simplest type of heat exchanger and consists of two concentric tubes. The sludge flows through the inner tube and the hot water through the outer concentric tube. The inner sludge tube is usually 100 mm in diameter which with recirculation pumping allows high velocities and freedom from blockage at reasonable heat transfer efficiency. It is easy to maintain and very reliable.

SPIRAL TUBE EXCHANGER. One of this type is also known as the Rosenblad heat exchanger. The exchanger tubes are spirally wound with the sludge and water countercurrent as in the concentric tube. The casing is circular with flat sides. The sludge flows in a curved path. Turbulence, selfsurface cleaning and thus heat exchange efficiency tends to be higher than with the concentric tube. It is however more expensive and more difficult to maintain than the concentric tube system.

(b) *Heat transfer and efficiency.* Concentric tube exchangers are regarded as offering high reliability and predictable performance. They are expensive because of the pipework, pumps controls and building necessary to house the system. Running costs are also high because of the continuous pumping through the heat exchanger.

The most important constraint on heat transfer efficiency of all types of exchanger is the upper temperature limit of the water circuit. Temperatures above 60–70°C can cause baking or fouling of the exchange surface. Typically therefore 300–350 J m^2/h is used for design purposes.

The heat exchanger system needs to be flexible with both variable water temperatures and working exchange areas. This will allow for differences in sludge characteristics and ageing of the equipment. This can often be achieved with modular units. An oversized exchanger could produce both overheating and fouling.

In order to maintain adequate performance then the exchanger must also be cleaned regularly to correct inevitable progressive fouling.

Data on overall heat losses from digesters is inconclusive (Brade *et al.*, 1982) but lagging digesters should be used particularly with the taller modern prefabricated tanks. This will reduce heat requirements, boiler rating and free digester product gas for other uses.

(c) *Steam injection*. The very high capital costs of the most reliable form of sludge digester heating (i.e. concentric tubes) led to investigations into cheaper and reliable alternatives. Two have been tried—direct steam injection and submerged combustion. Five steam prototype plants have now been built within the Severn–Trent Authority (Noone, 1984; Noone and Brade, 1985). Direct steam injection is an in-tank heating system, and no sludge pumping equipment or recirculating energy is required. The heat exchange surface (the steam itself) is self renewing. The equipment used consists of standard proven industrial items and is thus cheap and reliable. The most likely problem with direct steam injection was thought to be localised overheating which could result in process shocks or sterilisation of the biomass. Good mixing is essential (Brade *et al.*, 1984). The steam is delivered into the digester by a single lance.

All the pipework is lagged and inclined to condensate traps. The pipework also requires expansion sections and to be firmly anchored. Prior to entering the main the steam is filtered and derated through pressure reduction and control valves. The digester temperature is controlled by a modulating steam control valve. Early results from direct steam injection are excellent (Noone and Brade, 1985) although two long term problems were anticipated—corrosion and scale which could reduce plant life. Package water softening and conditioning equipment prevent scale and are necessary to avoid boiler damage. Corrosion has not been a major problem. There is however no background experience of steam-generating systems within the Water Authorities and special training and safety procedures are necessary.

(d) *Direct combustion*. Submerged combustion developed in a project at Severn–Trent but there are now other published data from the Yorkshire Water Authority (Hudson *et al.*, 1988).

Raw sludge is circulated through the burner zone of an open flame in a separate combustion tank prior to the digester. Special equipment is required to avoid corrosion, pump the sludge and control temperature (Kidson and Ray, 1984). Increased needs for pasteurisation may increase the use of this (or indeed the steam) heating option.

4.4.3 Sludge thickening

(a) *Sludge consolidation.* Most of the gas production and mixing performance of the digester can be equated to solids loading. A high solids concentration in the feed sludge will extend retention time, a low solids concentration can lead to a very short retention time and methanogen wash out. Digesters working at a thicker solids content are smaller and easier to heat and mix. The costs of the ultimate disposal of sludge by tanker are also reduced (Fig. 4.4). The characteristics of sludges vary and there is no simple solution to sludge thickening. There are therefore a variety of thickening processes such as centrifuges, chemical flocculation, belt presses and filters. In general, however, sludge consolidation by gravity has proved to be the most cost effective solution to thicken mixtures of primary and biological sludges (Brade *et al.*, 1984; Hoyland and Day, 1986). Sludge consolidation is a simple process relying on density difference between sludge solids and water. This conceals a complex set of principles and processes in operating 'simple' settlement. In sludge consolidation the sludge compresses by its own solids weight and water is squeezed out and upwards to form the supernatant. Sludges vary in their consolidation behaviour according to their surface properties and bound water. Sludge also ages because of microbial activity and its characteristics then change. Stirring is essential to free water and any gas generated in the detention period of the thickening process. Sludge consolidation is so complex that empirical test work is called for before the final ideal design (Hoyland and Day, 1986). In practice there is an optimum retention time of about 2 days. Sludge can be retained longer without too much gassing but often little further thickening takes place. If the retention approaches 5 days then there is significant deterioration in performance. The solids content of consolidated sludge is often improved by taller tanks. In shallow tanks the rate of consolidation is high but because the path length of water escaping is low and the weight of compressing sludge low then the solids content in the final sludge is also low. Conversely the rate of consolidation in tall tanks is low but the ultimate solids concentration is often high.

A typical working depth is 3–4 m. A proper picket fence stirrer is an essential part of the thickener. The stirring releases gas and water and promotes flocculation. An average rotational speed would be 0·1–1·0 rpm with bar spacings 300–400 mm and 50–60 mm diameter pickets.

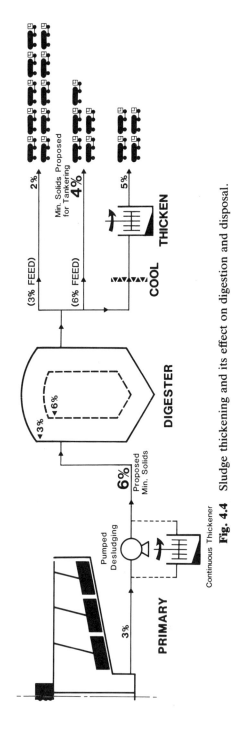

Fig. 4.4 Sludge thickening and its effect on digestion and disposal.

(b) *Sludge consolidation within primary tanks.* Sludge consolidation in primary tanks can be improved by controlling the rate of desludging. Ideally desludging should take place continuously with suitable pumps (peristaltic or double diaphragm). In some experiments an average solids content of 6% was achieved in this way (Brade *et al.*, 1984) without any deterioration in tank effluent performance. Separate scum removal from the primary tanks was necessary. Continuous desludging of the primary tanks enabled trials to be carried out on the benefits of continuous sludge feeding.

4.4.4 *Sludge feeding*

There are a number of benefits from continuous sludge feeding to anaerobic digesters. The main advantage is that it avoids sudden changes in organic load and shocks on the system but it also has beneficial effects on the in tank mixing and maintaining a stable temperature. Large slugs of cold raw sludge cause thermal stratification and short circuiting (Noone and Brade, 1982). Suitable pumps, which will avoid blockages while maintaining a low volumetric throughput, have not been available. The recent introduction of large peristaltics, double diaphragm and disc pumps seems to have overcome this problem (Noone and Brade, 1982, 1985).

An additional benefit from continuous feeding is a smoothing of gas production. Normally batch digester feeding causes large surges in gas production. Continuous feeding should allow a reduction in the size of expensive gas storage particularly where the gas usage is at an average. This effective gas storage as 'liquid' sludge is both safer and much less expensive than storage as the product gas.

One difficulty with continuous feeding is that the necessary equipment could be quite complex. The sludge should be screened prior to the digester and the pipework carefully designed to minimise blockage. Thus an apparent improvement in security of operation and reduction in labour might be replaced by more expensive complex maintenance.

4.4.5 *Gas storage*

Normally some form of gas storage is necessary to balance supply and demand. Traditional concrete digesters can be fitted with floating covers which rise and fall with production and use. The cover requires guiderails and safety equipment to avoid the introduction of air and explosive mixtures. Limit switches, and pressure and vacuum relief valves are also necessary to prevent damage to the bell. Separate gas-holders are required with prefabricated digesters, similar safety precautions are necessary together with antifreeze for cold weather. Gas flares are sometimes provided on stand-by in the event of an excess of gas. All the gas handling equipment should be made of corrosion-resistant materials and equipped with flame traps.

4.5 The digester

Traditionally Water Authority equipment was designed for a long life. Typically assets were written off over 30–60 years. This is entirely appropriate for reservoirs and many civil structures associated with domestic sewage treatment. Biological filters, for example, have a life expectancy of 50–75 years. Circumstances have changed, however, and the technology of mechanical and electrical process equipment is improving rapidly. In many cases, this sort of equipment would be obsolete in 3–5 years. Changes in the way central government finances the water industry coupled with privatisation means that process plant and equipment must now be treated in a similar way with a return on investment spread over a plant life of say 5–10 years. Conventional sludge digesters have been built of reinforced concrete and the major cost of sludge digestion was associated with the civil engineering of these reactors. It was apparent within the Severn–Trent Water Authority that the current and future requirements for sludge treatment could not be met by conventional sludge digestion while remaining within the new financial constraints (see Section 4.3). The historic (1976) capital cost of digestion, about £20–30 per head (Fig. 4.5) was too high.

The advantages of digestion for deodorisation and disinfection over other techniques cheaper in capital, such as pressing and filtering were, however, overwhelming and methods of improving process were investigated.

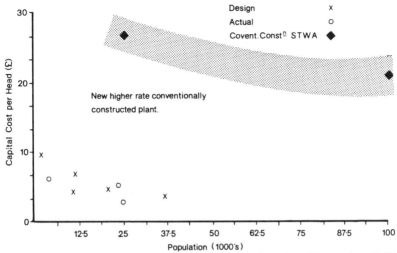

Fig. 4.5 Unit capital costs of prefabricated digestion plant (September 1982). (From Noone et al. (1984), reproduced by permission of the publishers Ellis Horwood Ltd.)

4.5.1 Retention time

The 1969 WPRL survey (Swanwick et al., 1969) and the Severn–Trent Survey (Brade and Noone, 1981) revealed that sludge digestion actually worked better at lower retention times (see Section 4.3). However, many of the digesters were designed with 20–30 days retention based on a 1954 report (Ministry of Housing and Local Government, 1954). A further difficulty was ignorance of the basic biochemical processes taking place during digestion. A fundamental understanding is necessary to diagnose and correct problems. Much of what is known about the microbiology and biochemistry of anaerobic digestion has been unravelled in the last 10 years (see Chapter 2). From an analysis of published laboratory and large scale industrial digesters critical retention time to avoid wash out can be established at 5 days (Henze and Harremoes, 1983; Stronach et al., 1986). This, however, assumes perfect mixing, 35°C temperature and a mostly soluble substrate. In practice some safety margin is required depending on substrate and temperature. Typically, for municipal sludge this is a factor of two (Parkin and Owen, 1986). There is now extensive operating experience within the Severn–Trent Water Authority at retention times between 10–15 days which show that there is little

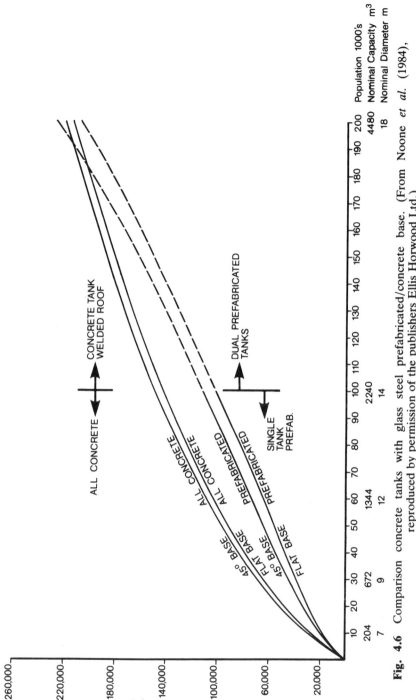

Fig. 4.6 Comparison concrete tanks with glass steel prefabricated/concrete base. (From Noone *et al.* (1984), reproduced by permission of the publishers Ellis Horwood Ltd.)

improvement in process performance above 10 days (Brade and Noone, 1981; Noone and Brade, 1982, 1985). Reducing the retention time of digesters from the conventional 30 days to 15 days would save between 30 and 50% of the capital costs depending on the size of the digester (Fig. 4.6) (Wolinski, 1984). The combined effects of improved ancillary equipment, e.g. sludge thickening and reduced detention times have now resulted in much smaller reactor volumes—typically a reduction of 25% of previous volumes has been achieved.

4.5.2 Prefabrication

A further conclusion from the Severn–Trent investigations, was that savings could be made by adopting prefabricated tanks and standardised modular designs. Such equipment was available in the farm and food industries where they had been in use for many years as storage silos. These tanks had already been adapted and used as digesters and settlement tanks in the food and farming industries. The tanks are made from enamelled mild steel plates bolted together (Fig. 4.7). The plates are manufactured off-site and thus on-site work is reduced to a minimum. The plate joints are sealed with mastic and grouted into a concrete base. The tanks require insulation (50–100 mm) and this can be internal, sprayed polyurethane foam, or external with a rockwool suitably weather-proofed with aluminium cladding.

A possible disadvantage of this construction could be that the largest single reactor size possible to date using this method within the Severn–Trent Water Authority is about $1000\,m^3$ (15 m high, 9 m diameter). This maximum has been set by the permissible size of the integral roof system also made from glass-lined steel. A plant contract for some 100 000 population has recently been awarded by another authority and it is understood that this will utilise individual tanks of $1800\,m^3$ volume (13·7 m high, 12 m diameter). Multiple tank installations could provide greater capacity and would also have advantages in respect of planned digester maintenance, i.e. sludge could still be treated while single digesters were overhauled. The modular approach would also avoid the need to initially oversize digesters to cater for anticipated expansions in flow (Thompson and Michaelson, 1984). All circumstances cannot be foreseen and a recent example case has arisen within Severn–Trent where some originally oversized digesters with at least 30 years remaining asset life can no longer

Fig. 4.7 Typical prefabricated enamelled steel tanks. (From Noone *et al.* (1984), reproduced by permission of the publishers Ellis Horwood Ltd.)

be used because of changes in local development plans and resewering. Previous development work (Noone and Brade, 1982) has shown that 500–600 m^3 volumes or population equivalents of 25–30 000 gives an optimum size for providing process equipment.

Other materials of construction have been considered and might be suitable in special circumstances. Ordinary mild steel plates could be used but would require weather-proofing externally and also require protection on-site during construction. There is unlikely to be any cost saving except in the unlikely case where a whole tank could be completed off-site. Similarly plastics or glass reinforced plastics could be used for smaller tanks (up to 100 m^3). Such materials have already been used for small farm and industrial digesters (see Chapters 3 and 5).

The new financial controls have also led to changes in procurement procedures within Water Authorities. Traditionally Water Authorities have issued detail plant specifications. This has often resulted in over-design and no risk sharing with the contractor. The alternative 'turn key' approach based on performance specification has enabled a few contractors to offer standardised and thus cheaper modular designs. The process risk is then borne by the contractor.

4.5.3 Aspect ratio

Tall reactors favour good mixing (see Section 4.4.1) and high aspect ratios (height to breadth) are to be preferred. Most northern European digesters are eggshaped for this reason. A major difficulty with high aspect ratio concrete digesters is their capital cost. Tall tanks increase construction costs and soil loading factors. Ratios above 1·5:1 rarely appear practical except by using the new lightweight materials outlined here (Thompson and Michaelson, 1984).

4.6 The residues after digestion

Some form of secondary storage after digestion is always required. Sludge disposal to land or sea may be hampered by weather or other transport difficulties and balancing or a storage buffer is essential. Sludge consolidation and thickening occurs during this type of storage and can also be used to reduce tankering costs. Transport costs are about £2/m^3. The overall charge could be reduced by half by increasing sludge solids from 2·5% to 5·0%. Unfortunately 'digested' sludge does not thicken easily and special precautions are required in the design of secondary thickening tanks. One of the major difficulties is continued gas production. Fresh digested sludge is still very warm and, with thermal stratification, can continue to produce gas for weeks after its introduction into the secondary tank. A second problem is flotation, with microbubbles reforming from the ongoing digestion and the water 'supersaturated' with methane and carbon dioxide. These all generate buoyancy and flotation of the sludge. Cooling and agitation is necessary to release these gases. Heat losses are much greater from the surface of most tanks than from the walls. Shallow tanks such as lagoons are much more effective thickeners than tall tanks. Approximately 10 days' retention per metre depth is required to achieve a 50% increase in solids concentration (Hurley et al., 1984). There are, however, often difficulties in finding sufficient land to enable shallow tanks to be built and 2–3 m depth is the most common. Final effluent may also be recycled into the secondary digestion tanks—this is called the elutriation process. This helps with rapid cooling and also provides some agitation to avoid flotation but this increases the size of the tanks. Gentle agitation can also be induced by picket fence scrapers which also have the advantage of sweeping consolidated sludge to a central hopper. Tanks of this design with a 30 m diameter are reported to work well (IWPC, 1979). Other combined techniques of cooling and agitation have been tried. Brade et al. (1984) report on heat exchangers and falling film coolers. Hurley et al. (1984) report on using aeration for cooling and thickening. Current practice is to provide 10–15 days' retention in a shallow circular tank (2–3 m deep, 30 m diameter) fitted with a picket fence thickener (IWPC, 1979).

4.7 The routine operation

Anaerobic digestion, like all biological processes, is inherently robust. The inertia in the system even at low detention times, i.e. 10–15 days is usually sufficient protection except where loading varies widely or process equipment is unreliable. Digesters are also simple to operate and do not need continuous monitoring. There have been temptations to complicate operation by installing sophisticated control equipment. However, complete automation can only occur when suitable sensors have been developed. Unfortunately this, along with most process areas, is not the case in much of sewage treatment today.

4.7.1 Control and monitoring

There are only three parameters worthy of continuous monitoring: temperature, gas production and sludge quality. Temperature is likely to vary a little according to ambient conditions and sludge feeding. It is usually the main process control loop, and used to suspend sludge feeding in the event of significant temperature drop. Gas analysis is the next most useful indicator. There should be a constant production of methane from an equivalent amount of organic matter metabolised (i.e. $0.5 \, m^3$ CH_4/kg volatile solids (VS) removed). Methane production is sensitive to normal variations in operation such as periodic feeding. This can be overcome by establishing an average mass balance and continuous monitoring based on a moving average. The methane content of the gas is typically 60–70%. This concentration can be monitored by calorimetry, infra-red spectroscopy or thermal conductivity.

Deviation from the norm is an early indication of problems. Calorimetry is currently the most reliable system, it simply burns the digester gas and records any temperature changes from a bimetallic strip in the flame. Gas volume is usually measured by turbine or diaphragm gas meters. Good results are also reported from the new vortex flow meters.

4.7.2 Analysis

(a) *Reduction of organic matter.* Traditionally removal or destruction of organic matter has been the most widely used performance indicator for anaerobic digestion. It is usually expressed as a percentage of the original organic matter in the raw sludge. The calculation has become commonly known as the van Kleeck formula (IWPC, 1979). Work by Noone and Brade (1982, 1985) however shows considerable day to day variations in the value of this parameter. The apparent volatile matter (VM) reduction was found to be favourably biased by precipitation of grit in the digester and there were also major difficulties in obtaining representative samples. Using simple total gas production as a control (without gas analysis) could also be biased by large amounts of carbon dioxide in a failing digester.

(b) *pH, alkalinity and carboxylic acids.* These parameters are closely linked and each gives some indications of digester performance. The process of anaerobic digestion relies on the production of carboxylic acids from solid organic matter and then the removal of these acids by methanogenic bacteria (see Chapter 2). Overall process difficulties arise from the difference in growth rates of the two groups of bacteria. The acidogenic bacteria respond to changes in organic matter more quickly than the methanogens.

pH. The pH optima of the methanogens is close to 7·0 and an imbalance between the two biochemical stages results in acid accumulation. pH is a logarithmic scale and is thus not on its own sufficiently sensitive to control digestion. Usually the detectable depression in pH comes *after* inhibition of the methanogens. Despite this the pH of the digester should be checked frequently since this will give an indication of steady state 'normal' values. Any deviation from this value will indicate biochemical changes within the system and should be regarded as an adverse sign. Continuous recording of pH is impossible because most electrodes are rapidly poisoned by H_2S and similar compounds.

CARBOXYLIC ACIDS. Normally the concentrations of carboxylic acids in a well operating digester are very low, <100 mg/litre. High concentrations of ionised carboxylic acids can be tolerated assuming there is sufficient buffering to prevent the formation of free acids and a shift in pH. Concentrations up to 3000 mg/litre do not cause inhibition assuming the pH is above 6·8. In a stable digester acetic acid is the principal carboxylic acid but as the digester is stressed the concentrations of butyric and propionic acid increase. This is a good indicator of difficulties but there is no easy method of analysing carboxylic acids without complex laboratory instrumentation.

ALKALINITY. The degree of acidity and therefore the stability of the digester can also be estimated from the alkalinity. Total alkalinity is measured by titration with a standard acid to pH 4·0. The problem with this simple total alkalinity determination is that any constituent which reacts with acid will be included. These typically are organic acids, sulphides and free ammonia. It is possible by titrating to various pH values to assess the contribution made by each ionic species. Titration of the sample to pH 5·75 results in a measure of the alkalinity due to bicarbonate, this is referred to as Partial Alkalinity (PA). Titration from 5·75–4·3 gives the alkalinity attributable to the volatile acid alkalinity, this is called Intermediate Alkalinity or IA (Ripley et al., 1985). The successful operation of a digester depends on maintaining adequate bicarbonate buffering and avoidance of excess acid so the ratio of IA/PA should give a simple and reliable indication of stability. The Environmental Protection Agency (1979) manual recommends a IA/PA ratio of between 0·1–0·3 for good operation. Changes in the alkalinity will also be reflected in the gas composition. The relatively simple acids/alkalinity ratio (IA/PA) and its consistency is the best practical process indicator found to date from the author's experience. At pH 7·0, with a bicarbonate alkalinity of 2000 mg/litre, the biogas will contain 30% CO_2. Sampling errors and fluctuations in operation from discrete analysis make interpretation difficult but continuous on-line monitoring of CO_2 should provide a possibility of monitoring both short and long term changes in digester stability in the future.

4.7.3 Effective biomass

Traditional and classical microbiological techniques of culture and most probable numbers (MPN) are not reliable but some indirect methods are available. One of the easiest to apply is to measure the methane generation from a standard substrate usually calcium acetate. Gas production is recorded with and without the unknown substance. This can give an indication of microbial activity and show the presence of any potential inhibitors (Speece, 1988).

4.7.4 Inhibitors

Complete process failure is now very rare and a wide range of synthetic organic materials can be degraded anaerobically. Only a few compounds lignin, n-alkanes, plastics and large organics with ether-type linkages seem to be resistant to breakdown. A number of

compounds can cause inhibition:
(1) heavy metals
(2) chlorinated solvents
(3) biocides and pesticides
(4) antibiotics

Modern trade effluent policies now make any process inhibitions by metals and detergents unlikely. Chlorinated solvents and special biocides remain potent inhibitors at very low concentrations, i.e. 1–10 mg/litre. They are also quite difficult to detect and any investigations into reductions in digester performance should include some test work in these areas.

4.8 Design and procurement procedures

The basic steps and check list to digester design are:

(a) Establish the sludge solids level from the primary and secondary settlement tanks including sludges from other works (it is prudent to check that there are no other treatment works difficulties limiting primary solids concentration, e.g. gravity sludge mains, tank scraper mechanisms, etc.).

(b) Consider the installation of a pre-digestion sludge thickening facility based on the raw sludge data from (a).

(c) Establish reactor volume from the quantity of sludge and a chosen detention period (12–15 days)—detention period is dependent upon the consistency of the solids concentration available for digestion.

(d) Define digester shape, i.e. aspect ratio (make as tall as economically and environmentally possible). If a tall and narrow digester is to be used, then a 45° (min.) conical base is possible. Some grit deposition is probably inevitable at 'economic' mixing power and any grit can then be removed with such a steep cone.

(e) Suitable structural/materials design choice should be arrived at by cost comparisons. Modular designs are frequently cheaper than large single structures for equipment, tanks and buildings.

(f) Secondary digester sludge tanks should be provided to promote cooling by: large surface areas, low aspect ratios, with steel or heat conducting sides. An alternative is a specific sludge cooler, with a thickener and taller storage tank.

(g) Mixing should be variable, external mixing is recommended.
(h) Heating should also be flexible; modular concentric tube systems are the most robust. Direct steam heating is currently under test and should be cheaper.
(i) Continuous feeding is ideal but pipework and pumps must satisfy the opposing requirements of large pipe capacity (to avoid blockage) and low pumping capacity (to facilitate continuous feeding). A recent development which has proved very successful within Severn–Trent Water Authority has used (110 mm diameter) peristaltic, double disc or lobe pumps to achieve reliable, low rates of raw sludge feed.
(j) Standby: all vulnerable equipment (usually that with moving parts) needs fitted or shelf standby.

4.9 Future trends

Much of the work described in this chapter has been concerned with reductions in the capital costs of anaerobic digesters, both within the process and by pre-thickening. Further reductions in capital costs may be envisaged with further standardisation and modular designs. Continuous feeding of digesters is also the key to using smaller gas-holders. Process equipment developments, such as the use of steam for digester heating, will continue to bring about savings in process equipment. Whilst alternative reactor configurations such as those described in Chapter 5 have been proposed, the solid components of sewage sludge and associated debris eliminate most options. Only subtle changes in tank geometry are envisaged as practical.

Thermophilic digestion ($>45°C$) is quicker and could reduce digester volume but they are still very unstable. The odours associated with thermophilic digestion liquors make it unacceptable at the moment.

Developments are also expected to reduce running costs. Heat recovery to pretreat raw sludge could provide savings but considerable developments in equipment to accept domestic sludges would be required. Further advances in sludge thickening are also likely, the costs of tankering are certain to increase and additional post-digester thickening could further reduce these costs. Power generation may also prove to be a useful by-product particularly as the power industry is deregulated and privatised. Modular reliable, maintenance free, mass produced engines would be necessary.

4.10 Conclusions

The main change in anaerobic sludge treatment, in common with other areas of water and public health engineering, has been in overall process design. There has been a progression in design from an initial hydraulic 'storage' emphasis, followed by a period of dominance by structural design to finally the current emphasis on process design. These changes have in part been brought about by changes in Water Authority financing. The anaerobic digestion process is, and will continue to be, a very attractive method of domestic sludge treatment. The developments outlined in this chapter can reduce capital costs by between 30 and 50% compared to traditional designs.

The anaerobic digester is now considered as a biological reactor which must be kept at the right temperature and well mixed. Reliable process equipment is a vital part of achieving this objective. Future developments in ancillary equipment with reductions in operating costs are envisaged, but the greatest changes may be brought about by European legislation and guidelines governing the ultimate disposal of sewage sludge.

4.11 Acknowledgements

The author wishes to express his sincere thanks to his colleagues and the Severn–Trent Water Authority. Colin Brade his principal development collaborator is especially thanked. Any views expressed in this chapter are those of the author and do not necessarily reflect those of the Severn–Trent Water Authority.

4.12 References

Baumann, P. G. and Huibregtse, G. L. (1981). Evaluation and comparison of digester mixing systems. *J. Wat. Poll. Contr. Fed.*, **54**, 1194–1203

Brade, C. E. and Noone, G. P. (1981). Anaerobic sludge digestion—need it be expensive? I. Making more of existing resources. *Wat. Pollut. Control*, **80**(1), 70–90.

Brade, C. E., Noone, G. P., Powell, E., Rundell, H. and Whyley, J. (1982). The application of developments in sludge digestion within the Severn–Trent Water Authority. *Water Pollution Control*, **81**(2), 200–19.

Brade, C. E., Noone, G. P. and Whyley, J. (1984). Progress in anaerobic digestion—heating, cooling and thickening. In *Sewage Sludge Stabilisation and Disinfection*, ed. A. M. Bruce. Ellis Horwood, Chichester, pp. 158–73.

Carroll, W. D. and Ross, R. D. (1984). A full scale comparison of confined and unconfined gas lift mixing systems in anaerobic digesters. In *Sewage Sludge Stabilisation and Disinfection,* ed. A. M. Bruce. Ellis Horwood, Chichester, pp. 146–57.

Commission of the European Community (1981). *The Treatment and Use of Sewage Sludge.* COST Project 68 bis final report 56–58.

Department of Scientific and Industrial Research (1960). *Water Pollution Research. The Report of the Water Pollution Research Laboratory 1955– 1960. Annual Reports,* HMSO, London.

Environmental Protection Agency (1979). The disposal of sewage sludge. *Federal Register 1979,* **44,** 179, Washington, USA, pp. 53438–64.

Environmental Protection Agency (1979). *Process Design Manual for Sludge Treatment and Disposal.* EPA Centre for Environmental Research, Information Technology Transfer EPA 625/1-79-011 Washington, USA.

Farrell, J. B. (1984). Recent developments in sludge digestion in the United States and a view of the future. In *Sewage Sludge Stabilisation and Disinfection,* ed. A. M. Bruce. Ellis Horwood, Chichester, pp. 317–29.

Hall, E. R. and Hobson, P. N. (1988) (eds). *Anaerobic Digestion 1988,* Pergamon, Oxford.

Henze, M. and Harremoes, P. (1983). Anaerobic treatment of waste water in fixed film reactors: a review. *Wat. Science and Technology,* **15,** 1–90.

Hoyland, G. and Day, M. (1986). An evaluation of picket fences for assisting the consolidation of sewage sludge. *J. Wat. Poll. Contr.,* **85**(3), 291–303.

Hudson, J. A., Bruce, A. M., Oliver, B. T. and Auty, D. (1988). Operating experiences of sludge disinfection and stabilisation at Colburn Sewage Treatment Works, Yorkshire. *J. Institution of Water and Environmental Management,* **2**(4), 429–41.

Hurley, B. J. E., Rachwal, A. J. and Hatton, C. J. (1984). Consolidation of digested sludge. In *Sewage Sludge Stabilisation and Disinfection,* ed. A. M. Bruce. Ellis Horwood, Chichester, pp. 239–55.

Imhoff, K. R. (1984). The design and operation of anaerobic sludge digesters in Germany. In *Sewage Sludge Stabilisation and Disinfection,* ed. A. M. Bruce. Ellis Horwood, Chichester, pp. 201–23.

Institute of Water Pollution Control (1979). *Sewage sludge 1. Production, Preliminary Treatment and Digestion. Manuals of British Practice.* IWPC, Maidstone, Kent.

Kidson, R. J. and Ray, D. L. (1984). Pasteurisation by submerged combustion together with anaerobic digestion. In *Sewage Sludge Stabilisation and Disinfection,* ed. A. M. Bruce. Ellis Horwood, Chichester, pp. 399–408.

Ministry of Housing and Local Government (1954). *Report of an Informal Working Party on the Treatment and Disposal of Sewage Sludge.* HMSO, London.

Noone, G. P. (1984). Cost effective treatment and disposal of sludge to environmentally acceptable standards. *6th EWPCA Symposium,* Munich, 1984.

Noone, G. P. and Brade, C. E. (1982). Anaerobic sludge digestion—need it be expensive? II. Higher rate and prefabricated systems. *Wat. Poll. Control,* **81**(4), 479–510.

Noone, G. P. and Brade, C. E. (1985). Anaerobic sludge digestion—need it be expensive. III. Integrated and low cost digestion. *J. Inst. Wat. Poll. Contr.*, **84**(3), 309–28.

Noone, G. P., Brade, C. E. and Whyley, J. (1984). Progress in anaerobic digestion—prefabrication of digesters. *Sewage Sludge Stabilisation and Disinfection*, ed. A. M. Bruce. Ellis Horwood, Chichester, pp. 107–24.

Parkin, G. F. and Owen, W. F. (1986). Fundamentals of anaerobic digestion of waste water sludges. *J. Environmental Engineering*, **112**(5), 867–920.

Ripley, L. E., Boyle, W. C. and Converse, J. C. (1985). Improved alkalimetric monitoring for anaerobic digestion of poultry manure. *Proceedings of the 40th Industrial Waste Conference 1985*. Purdue, Indiana. Butterworth, Boston, pp. 141–9.

Rundle, H. and Whyley, J. (1981). Comparison of gas recirculation of systems for mixing in anaerobic digestion. *Wat. Pollut. Control*, **80**(4), 463–80.

Speece, R. E. (1988). A survey of municipal anaerobic sludge digesters and diagnostic activity assays. *Water Res.*, **22**(3), 365–72.

Stanbridge, H. H. (1976). *History of Sewage Treatment in Britain*. Institute of Water Pollution Control, Maidstone.

Stronach, S. M., Rudd, T. and Lester, J. N. (1986). *Anaerobic Digestion Processes in Industrial Waste Treatment*. Springer-Verlag, Berlin, 184 pp.

Swanwick, J. D., Shurben, D. G. and Jackson, S. (1969). A survey of the performance of sewage sludge digestion in Great Britain. *J. Inst. Wat. Poll. Control*, **68**(6), 639–61.

Thompson, J. L. and Michaelson, A. P. (1984). Design aspects of the new anaerobic digesters at Bury. In *Sewage Sludge Stabilisation and Disinfection*, ed. A. M. Bruce. Ellis Horwood, Chichester, pp. 92–106.

Vincent, A. J. and Critchley, R. F. (1984). A review of sewage sludge treatment and disposal in Europe. In *Sewage Sludge Stabilisation and Disinfection*, ed. A. M. Bruce. Ellis Horwood, Chichester, pp. 550–71.

Whitehead, H. C. and O'Shaughnessy, F. R. (1931). The treatment of sewage sludge by bacterial digestion. *Proceedings of the Institution of Civil Engineering*, 38–135.

Wolinski, W. K. (1984). A cost comparison of prefabricated and conventional digesters. In *Sewage Sludge Stabilisation and Disinfection*, ed. A. M. Bruce. Ellis Horwood, Chichester, pp. 488–99.

5 Anaerobic digestion: industrial waste treatment

A. D. Wheatley
Cranfield Institute of Technology, Bedford, UK

5.1	**Introduction**	172
5.2	**Types of waste**	176
5.2.1	Food industry wastes	176
	(a) The dairy industry	177
	(b) Starch/sugar and sugar confectionery	178
	(c) The sugar industry	178
	(d) Confectionery/soft drinks	178
	(e) Brewing	179
	(f) Distillery and fermentation industry	179
	(g) Vegetable processing	180
	(h) Meat processing	181
5.2.2	Other types of wastes	181
	(a) The paper industry	181
	(b) The chemical industry (including textiles and pharmaceuticals)	182
5.3	**The treatment of industrial wastes: general principles**	183
5.3.1	General principles	183
5.3.2	Avoiding waste	184
	(a) Control	184
	(b) Product formulation	184
	(c) Process modification	184
	(d) Stream separation	184
	(e) Equipment design	184
	(f) Recycling	184
	(g) By-products	184
5.3.3	Waste treatability	185
5.4	**Effluent treatment processes**	186
5.4.1	On-site treatment	186
	(a) Separation of solids waste	186
	(b) Consolidation	186
	(c) Effluent	186
5.4.2	Pretreatment processes	186
	(a) Screening	188
	(b) Grit removal	188

172   A. D. Wheatley

	(c) Flow balancing/two-stage treatment	188
	(d) Control	189
	(e) Settlement and flotation	190
5.4.3	The anaerobic reactor	191
	(a) The contact stirred tank reactor	193
	(b) The Upflow Anaerobic Sludge Blanket Reactor (UASB)	199
	(c) The fluidised bed	203
5.4.4	Materials of construction	204
5.5	**Ancillary equipment**	206
5.5.1	Heating	206
5.5.2	Gas storage	206
5.5.3	Biogas utilisation	206
5.5.4	Settlement	207
5.5.5	Aerobic treatment	207
5.6	**The control of digestion**	207
5.6.1	Safety	207
5.6.2	Instrumentation	208
	(a) Choice of sensors	209
	(b) System and software	210
5.6.3	Laboratory tests	210
	(a) Alkalinity	210
	(b) Biological tests	210
5.7	**Economics**	211
5.8	**Treatability problems**	214
5.8.1	Types of inhibition	214
5.8.2	Nutrient requirements	214
	(a) Trace nutrients	215
5.8.3	Toxic materials	216
	(a) Metals	216
	(b) Disinfectants and detergents	216
	(c) Solvents	216
	(d) Biocides and pesticides	216
	(e) Ammonia	217
	(f) Sulphur compounds	217
5.9	**Conclusions**	218
5.10	**References**	219

5.1 Introduction

The anaerobic treatment of industrial waste is a new more efficient technology which has developed rapidly in some European countries notably Holland, Belgium, France, Germany, Italy and Scandinavia, but less rapidly in others particularly the United Kingdom, Spain and Greece. The application of anaerobic processing for industrial waste is now a widely accepted technology. As the running costs are cheaper compared to the alternatives, it should be the treatment of choice for strong wastes. There are about 600 plants in Europe but only 15 so far in the UK (Table 5.1). There are three principal reasons for this relatively poor uptake of the new technology by UK industry:

(1) Waste disposal charges in Holland, Germany and Scandinavia, where most of the anaerobic equipment is installed, are significantly higher than in the UK.
(2) Many of the remaining anaerobic treatment plants in France and Italy have attracted generous National or European Commission (CEC) subsidies. In Italy, for example, anaerobic treatment plants can attract a 70% capital grant because the country is short of indigenous fossil fuel.
(3) There is, therefore, a larger market for waste treatment plant in mainland Europe and the companies selling the technology are large, profitable and able to provide the necessary research and technical assistance in the event of unforeseen problems. The UK industry is small and fragmented by comparison.

The way in which the protection of the environment in the UK is organised is undergoing a major reorganisation. The water industry has been privatised and a National Rivers Authority set up. Much of the new pollution legislature seems certain to arise from Europe. There has been increasing government pressure on Water Authorities over the last five years to improve the return on investments and generate surpluses. This has meant increases in water charges of about 15%. There have been substantial efficiency gains within Water Authorities but water quality and legislature demands are also increasing. Implementation of the European Directives on drinking, bathing and surface water together with the sludge disposal directive could cost the water industry between £5–20 billion (billion = 10^9). The recent announcement of the value of the 'k' factor, or the premium above the retail price index that Water Authorities will be able to charge, means waste charges to industry will rise at between 15–20% per annum (Anon., 1988). There is, as a consequence, a very strong interest in new and cheaper methods like anaerobic treatment

Table 5.1 UK anaerobic industrial plants

Owner	Waste	Location	Type of reactor	Contractor	Date commissioned
J. Sturge	Molasses	Selby	CSTR	Ames Crosta Esmil	1970 rebuilt 1988
Tenstar Products	Starch	Ashford	CSTR	Biomechanics	1976
British Sugar Corporation	Sugar beet	Bury St Edmunds Peterborough Lincoln	CSTR CSTR CSTR	Sorigona	1982 1986 1988
McCains	Potato	Peterborough	UASB/Lagoon	ADI	1981
Swizzels Matlow	Confectionery	Stockport	Filter	Prototype ETA	1982
Caernarvon Creameries	Dairy	Chwilog, Caernarvon	UASB	Hamworthy	1982
Distillers	Yeast production	Stirling	Filter	Biomass	1986
Davidson	Paper mill	Aberdeen	UASB	Pacques	1986
Wrexham Lager	Brewery	Wrexham	CSTR	Biomechanics	1987
Callard & Bowser	Confectionery	Bridgend	CSTR	CLEAR	1985
Hall & Woodhouse	Brewery	Blandford Forum	CSTR	CLEAR	1987
Cricket & Malberie	Dairy	Metherstowie, Somerset	CSTR	CLEAR	1987
General Foods	Starch/coffee	Banbury	UASB	Biwater	1988
Tunnel Refineries	Starch	London	CSTR	Biomechanics	1988
Coca Cola	Soft drinks	Wakefield	UASB	Esmil	1989

Table 5.2 Trade effluent changes 1987/88

Region	Average sewage strength		R P/m^3	V P/m^3	B P/m^3	S P/m^3
	COD mg/litre (Os)	SS mg/litre (Ss)				
Anglia[a]	680	400	5·3	6·3	9·6	2·5
Northumbrian	351	171	7·3	3·7	7·1	3·5
North West	363	258	4·0	3·4	4·2	2·6
Severn–Trent	351	322	5·6	5·9	6·1	3·5
Southern	452	512	7·6	8·7	10·2	7·0
South West	545	435	9·6	9·8	15·3	14·0
Thames	450	331	3·6	4·2	6·5	8·3
Welsh	500	350	3·8	3·0	7·0	4·9
Wessex	383	281	2·1	5·9	5·6	5·1
Yorkshire	943	315	—	8·7	9·2	5·6

$$C = R + V + B\frac{(Ot)}{(Os)} + S\frac{(St)^b}{(Ss)}$$

where C = total charge
 R = reception and conveyance charge
 V = volume and primary treatment charge
 Ot = COD of effluent after 1 h quiescent settlement at pH 7
 Os = COD of crude sewage after 1 h quiescent settlement
 B = biological oxidation cost/m^3 settled sewage
 St = suspended solids (mg/litre) of trade effluent at pH 7
 Ss = total suspended solids (mg/litre) of crude sewage
 S = treatment and disposal cost of primary sludge

[a] Anglia water now charge for COD + (4·5 × organic nitrogen) analysed after settlement because of local difficulties with nitrogen.

[b] The formula is based on complete treatment. It would be altered to remove the oxidation component if just settlement and screening were used. Simple screening and settlement are common forms of treatment prior to coastal discharges.

for industrial waste treatment. Table 5.2 shows the current method of calculating waste disposal to sewer changes (Micklewright, 1986). Some basic research and development is still required to improve reliability and speed start up, but at large factories industrial effluent treatment can already be applied to economic benefit. This chapter will discuss the currently available technology, the costs, the most treatable effluents and the most frequently encountered problems. It concludes with some of the requirements needed to improve the process in the future.

5.2 Types of waste

The largest industrial sector in Europe is food, drinks and fermentation (UK turnover £25B, Chemical Industry £20B) (Chemical Industries Association, 1986). All the industries in this sector are large water users. In the USA and Scandinavia the paper industry is the largest water user (British Paper and Board Federation, 1987). Most of the larger industrial anaerobic treatment plants built so far are in these two sectors.

5.2.1 Food industry wastes

The hygiene and quality control necessary in the food industry results in a high wastage. There are few opportunities for recycling because of the stringent cleanliness required. Product to water ratios above 10:1 are the norm (Table 5.3). There is a wide range within and between sectors but when overall size is taken into account then four industries can be identified as likely to benefit from anaerobic treatment.
1. Milk and milk products.
2. Starch products and sugar confectionery.
3. Brewing.
4. Distilling and fermentation.

Table 5.3 Strength and waste water generated in the food industry

Industry	Strength (typical) BOD mg/litre	Water to product ratio litres/unit of product
Dairy	500–750	12:1
Butter and cheese	1 500–2 500	3·0:1
Distillery	1 500–2 000	20:1
Brewery	500–1 500	8:1
Maltings	2 000–3 000	15:1
Food canning	100–1 300	10:1
Frozen peas	1 000–2 000	12:1
Chips and other vegetables	1 000–1 500	20:1
Chickens	600	30:1
Slaughterhouse	1 000–2 000	20:1
Sewage	250	130 per head

These industries also have the largest single sites and effluent disposal costs may be £100–300 000 per annum per factory (MAFF, 1986). Most of the other food processing sites are too small to enable economic effluent treatment. In the UK, for example, there are over 5000 food processing sites but only about 150–200 have loads greater than 250 tonnes of COD per annum, the point at which treatment becomes economically very attractive. These are the major markets for the application of industrial anaerobic treatment. In principle, food industry effluents should be easily treated. In practice there are a number of difficulties despite the fundamentally biodegradable constituents. The main problems are the high strength and fluctuations which occur in the type and quantity of wastes to be treated. The cleaning aids and sanitisers used create further problems. The fluctuations in flow can be 10–200% of average and the strength can vary from nothing in cooling water to a COD of several hundred thousand for accidentally lost product or raw materials. The high strength of the waste normally also makes them deficient in trace nutrients particularly nitrogen which reduces treatability (Section 5.8).

(a) *The dairy industry.* The dairy industry is currently the largest single source of industrial effluent in Europe. They also have the most rural sites so discharge to sewer is often impossible. The milk industry spends about 60MECU a year on effluent treatment, 50% of it on sludge disposal which could be reduced by anaerobic treatment. The industry is centred on France, Germany, Holland, Ireland and the UK. There are 10 full-scale anaerobic treatment plants in Europe, three in France, two in Holland, two in the UK, two in Ireland and one each in Germany and Spain. Compared to other industries this is a small number but is not thought to reflect any resistance to anaerobic digestion by the dairy industry. Most dairies had to install or refurbish effluent treatment in the period 1970–75 during an expansion of the industry. Much of this equipment is still serviceable. Since this period the dairy industry has undergone considerable restructuring and consolidation. Most dairy products—butter, cream, cheese and concentrated milk are now produced in large rural dairy complexes. A typical European dairy processes 500 m^3 of milk per day and generates a similar volume of effluent. The waste is from washing of vats, evaporators and transport. Highly alkaline cleaning materials are used which can raise effluent pH to toxic levels. Balancing and dump storage are essential. COD values of dairy effluents can vary from

100 to 2250 mg/litre, but those from dairy products are warm and strong, ideal for anaerobic treatment. Nine anaerobic treatment plants have been built in the period 1983–88. The low solids of dairy type effluents have favoured the use of filter configurations (see Section 5.4)—five out of the nine current plants.

(b) *Starch/sugar and sugar confectionery.* This has been one of the most successful areas of application of anaerobic digestion. Starch is used in snack foods, ready meals and desserts. It is made from maize, rice and wheat. The effluent from starch production is strong (15 000 COD), ideal for anaerobic digestion. A high proportion of the COD is colloidal solids. These solids reduce biodegradability and favour solids tolerant processes, i.e. UASB and Contact (see Section 5.4) if a single stage is to be used. Problems have been encountered with existing plants which are some of the oldest installed. Two-stage treatment with separate prehydrolysis of the solids is now an accepted method of overcoming problems associated with treatability. The anaerobic treatment of starch wastes is now an accepted technology and the market is expected to continue to grow.

(c) *The sugar industry.* The sugar industry is a major water user with consumption between 20 and 30 times the product produced. The industry has had to take steps to conserve water. Most processing factories now have extensive recirculation, and a concentrated effluent is produced. Much of the European sugar industry is rural and seasonal, this requires on-site effluent treatment. The effluent rapidly deteriorates generating acids which make it unsuitable for aerobic treatment. Lime is used to counteract the acidity and prevent malodour. The alkalinity and calcium in the lime aid anaerobic treatment. The biomass in the anaerobic digester flocculates around the lime and promotes settlement in the Contact Process (see Section 5.4). A disadvantage of lime is it accumulates during the season reducing the space in the digester for active biomass. Solids tolerant digesters, the Contact Process and UASB reactors are necessary (see Section 5.4). Anaerobic digestion followed by aerobic polishing before discharge to a water course is an ideal form of treatment. Anaerobic treatment is an accepted type of treatment for sugar wastes.

(d) *Confectionery/soft drinks.* Much of the waste from the confectionery industry is sugar-based and COD values can be large,

5–10 000 mg/litre. The wastes are from cleaning and they are periodic and usually hot. Both aerobic and anaerobic treatability are low because of acidity and lack of additional nutrients. The costs of correction have reduced the use of on-site treatment. The strength of the effluents however means that the costs of discharge to sewer are also high and three anaerobic treatment plants have been built in Europe for confectionery waste.

(e) *Brewing*. Europe's breweries are thought to spend 60MECU on waste disposal each year. Like the dairy industry production facilities are large but unlike the dairy industry most breweries are urban. They therefore have the option to discharge wastes to mains drainage. There is extensive recovery of by-products from brewing, e.g. malt residues as animal feed, spent hops as fertiliser, yeast as food and CO_2. Wash waters are weaker than the dairy industry. Water consumption is between 5–20 litres per litre of beer produced.

COD values are within the range 500–1500 mg/litre with a neutral pH. Brewery wastes are easily treated aerobically and anaerobic treatment is less attractive than for some other food wastes because the effluent is weaker and colder. The UK brewery industry has been very active in testing and researching anaerobic treatment as a method of pretreatment prior to discharge to sewer. At present a number of full-scale applications have been delayed until after privatisation of the water industry. Increased water supply and effluent disposal charges will encourage more efficient water usage. This will increase effluent strength and improve the energy balance for anaerobic digestion. Research is progressing on lower temperature digestion. Small anaerobic digesters could be utilised for the strong warm process streams ignoring most of the cool wash waters.

(f) *Distillery and fermentation industry*. Distillery wastes are frequently very strong (COD values 10–60 000 mg/litre, solids 10 000 mg/litre). Like the brewery industry recovery of by-products animal feeds, fertiliser and CO_2 is a traditional part of distilling. Water consumption is on average 10:1 (water to product) but volumes are smaller than either the dairy or brewery industries. Many distilleries therefore particularly in the UK, where they also are small and rural, have been able to dispose of effluents by land irrigation or discharge to sea. A small Scottish highland distillery, for example, will produce 500 m^3 of effluent a week. Large industrialised distilleries, however,

Table 5.4 An example of a typical anaerobic treatment of wine distillery effluent

Reactor type—downflow anaerobic filter
Reactor size—5 600 m^3
Loading rate—11·8 kg COD m^3/day
Daily load—66 tonnes of COD
COD removal efficiency—90%
Effluent flow rate—70 m^3/h
Biogas use—boiler-electricity generation
Capital cost—FF1·8 million
Location—Recivo (Martell/Hennessy/Courvoisier Group) Cognac, France
Contractor—SGN (France)

(Camilleri, 1988)

have major waste disposal problems. One French cognac producer, with an anaerobic digester, generates 10 000 m^3 of effluent a week (Table 5.4), equivalent to the waste from a population of 0·5 million. These wastes have previously been treated aerobically but only after dilution with other wash waters and recycling. Aerobic waste treatment plants have always been difficult to operate because of the acidity of the waste, high temperatures and high oxygen demands. Thus distillery wastes are ideal for anaerobic treatment, there are few alternatives, and the sector has the largest number of installed plants, 25, (1983–1988) most in Italy and France. Many plants are contact stirred tank reactors (CSTRs) but there are a significant number of filters in France.

(g) *Vegetable processing.* The vegetable and fruit processing industry like the dairies and sugar industry are rural. They may like the sugar industry also be seasonal although many process different crops at different times of the year. Effluents are, however, dilute and cold and this has favoured aerobic treatment. There are two exceptions, potatoes and peas, both release easily soluble starches and produce a higher strength waste. Anaerobic digesters have been built at 11 potato and potato snack factories. They are mostly the sludge blanket

UASB type or CSTR (see Section 5.4) to cope with the high concentrations of colloidal solids from potatoes.

Pea processing wastes are also strong (COD 4000 mg/litre) and could be treated anaerobically but the season is so short—10–12 weeks on average—that the capital costs of anaerobic treatment compared to aerobic are uncompetitive. There are at present four non-potato vegetable processing plants in Europe (three filters—see Sections 5.4); this number is expected to grow at the rate of 1–2 per year.

(h) *Meat processing.* Meat processing wastes are significantly different from other food industry wastes. They are very strong containing grease, blood and faeces. They contain significant amounts of recalcitrant organic matter, such as straw and hair.

The wastes frequently include high concentrations of biocides and disinfectants such as hypochlorite. There are also problems arising from malodours and low temperatures. These are, therefore, very difficult wastes for any form of treatment and better separation of the various process streams arising from meat processing is the best form of waste management (see Section 5.3). The difficult nature of the waste has encouraged anaerobic trials, test work and demonstration plant. Experiments in the UK and USA have been carried out since 1959. Only solids-tolerant long retention time reactors (i.e. CSTR—see Section 5.4) are likely to succeed.

5.2.2 *Other types of waste*

(a) *The paper industry.* The paper industry is a very large water user, a tonne of paper or board requires 40 tonnes of water. The potential pollution and costs of such a high water usage means that recycling has been essential for some time. The loss of fibres in waste can account for 1·5–2% of mill production, recovery and water re-use are therefore attractive. Losses have in most cases been reduced to 2:1. Doubts about the resilience of anaerobic compared to aerobic effluent treatment of different industrial waste has meant that there was little interest in the technology when anaerobic digestion was being developed by the food industry. Successful trials in Scandinavia in 1980–83 led to a very rapid dissemination and take up of the technology world-wide. Fourteen plants with an investment of 24MECU have been installed since 1983.

Paper is prepared either mechanically (mechanical/fibrewood) or chemically using the kraft process. Wood and wood pulp can also be substituted by grass, straw, rags and waste paper. Raw paper is then bleached with sulphite waters (see Pearson, 1988, for a review). There are major problems for any form of treatment including detergent and sulphur toxicity; recalcitrant fibres (lignin), inorganic fillers and pesticides or other wood treatments. Rates of biodegradability are low. The wastes are strong and warm, average temperature 35°C, COD 10–20 000 mg/litre. This favours anaerobic treatment despite the difficulties. Stream separation, pilot work and resilient designs are necessary. There is still much ongoing research in the major pulp-producing regions and the number of anaerobic digesters in this area world-wide will continue to increase. Europe is not, however, an important pulp producer or processor. These are Scandinavia, Canada, the USA and Japan.

(b) *The chemical industry (including textiles and pharmaceuticals).* Generally if a chemical processing waste has been found to be treatable aerobically, then it will also be treatable anaerobically. The main advantage of anaerobic treatment is the low sludge production, but this becomes a disadvantage during start up, shock toxic discharges or in the event of a large loss in biomass. The application of the technology parallels that in the paper industry. Originally it was assumed that treatment would be difficult but it has now been demonstrated that the highly reducing conditions in an anaerobic digester may lead to the improved biodegradation of certain compounds, e.g. colour and chlorinated organics (Battersby, 1989). There are, however, four common components of chemical wastes which interfere with the treatment (see Section 5.8):

(1) sulphates
(2) chlorinated solvents
(3) surface active agents
(4) recalcitrants

There is currently much laboratory and pilot research on the anaerobic treatment of recalcitrant and hazardous wastes. Nutrient or chemical supplementation will be necessary. Acclimatisation and reversibility of toxicity has been shown to be possible with anaerobic cultures. There are few full-scale anaerobic treatment plants dealing with chemical wastes. There are three in the USA on celanese synthetic textile waste and three to four on pharmaceutical wastes. The pharmaceutical wastes mostly resemble molasses distillery effluents. Much of the European chemical industry is coastal or Rhine-based and has not so far been subject to the same controls as the inland food and paper industry. As restrictions on the pollution of European seas increase then there will be significant spending on pollution control equipment by the chemical/petrochemical industries. Stream separation, analysis and treatability tests will be an important part of these applications. Effluent treatment plants will be combinations of physico-chemical treatment, anaerobic and aerobic units. Recovery and recycle as in the paper industry will be essential.

5.3 The treatment of industrial wastes: general principles

5.3.1 General principles

In most industries there is still much scope for reducing waste or using very simple physical treatments, such as screening, separation and settlement to reduce the cost of waste disposal. This chapter focusses on one of the most complete technologies available for treatment. Anaerobic digestion can reduce effluent disposal costs by 80%. Like most bioprocesses, however, anaerobic treatment depends as much on upstream and downstream equipment as it does on the fermenter step itself. All total effluent processes are quite complex and their apparent cost effectiveness needs to be carefully assessed to include factors for reliability, maintenance and manning. A thorough understanding of the production processes and wastes is necessary to assess possible savings. This may involve flow and chemical analysis. Continuous monitoring of the strength of a waste is usually impossible for most production processes and a sampling survey is needed. The survey is based on a typical operating cycle of the unit process under test. Samples can be taken on a time weighted average or more realistically on a volume proportional basis. One of the aims of survey work is to compile a mass balance of raw materials, product and waste. From the information gathered it is possible to detail:

(1) Losses of product and raw materials from each unit process.
(2) The cost of disposing of this waste.
(3) The cost of wastage as a percentage of value added.
(4) The selection and cost of treatment alternatives.
(5) Possible control procedures.

5.3.2 Avoiding waste

Once the sources and quantities of waste have been established it is then usual to find that there are potential methods for improving process efficiency. There is a basic check list to follow to avoid waste:

(a) *Control.* The aim of better control is to deliver precise amounts of raw materials, reactants, catalyst, physical change and water at exactly the right time. This avoids wasting energy, reactants and water.

(b) *Product formulation.* The objective of a change in product formulation would be to use alternative cleaner raw materials which could reduce the waste produced without altering the quality of the product.

(c) *Process modification.* Similarly alternative processing operations might also be used, such as dry processing. This could be possible without affecting efficiency and quality.

(d) *Stream separation.* Charges are based on volume and strength so it is often possible to separate and divert very strong streams, i.e. fat, slurries, etc., to avoid contamination of the main stream. These may then be disposed of differently.

(e) *Equipment design.* A change in the design of the existing machinery might also be considered to reduce pollution. Possibilities are a lower operating temperature, using less water, and preventing mechanical damage again without interfering with quality and efficiency.

(f) *Recycling.* The most common method of recycling is to incorporate damaged raw material or product back into the process. This can frequently be done without adverse effects on quality. A second possibility is to recycle a high grade water from cooling or final product washing, for a lower grade use such as washing raw materials.

(g) *By-products.* This is already widely practised and can significantly offset disposal costs. The simplest approaches are:

(1) Recovery of animal feeds.
(2) Fertilisers.
(3) Waste paper or card.
(4) Energy by combustion of waste.
(5) Fats and oils.

More complex by-products involving further processing may also be justified:

(6) Starch recovery for alternative reconstituted starch products, e.g. from potato crisp products.
(7) Protein for alternative protein products, e.g. Bovril, Marmite, sauce additives.
(8) Protein production (SCP) from waste carbohydrates, e.g. Mycoprotein.
(9) High value added by-products such as special oils, carbohydrates and proteins, vitamins, flavours, fragrances, colours and fermentation aids for the food, pharmaceutical and fermentation industry. The recovery of high value by-products is still very speculative and is the subject of further research.

5.3.3 Waste treatability

For some wastes treatability data are available for the design of anaerobic treatment plant without intermediate laboratory and pilot studies. This is true of domestic sludge and some food wastes applications of the technology where suppliers have already installed such plant at several similar factories using common design principles. There are still many wastes for which anaerobic treatability data are not available or where the processing operations are significantly different.

The most frequently encountered problems with industrial wastes compared to domestic wastes are a lack of nutrients, variability and persistent toxins. Many of these difficulties can be predicted and overcome by laboratory treatability trials before design. Such tests are essential if there are any doubts about the characteristics of the waste. A laboratory treatability study will identify any reliability problems which were not apparent from the initial single analysis as well as comparing competing reactor types and likely process loading rates. Expenditure on this type of work is less than 0·5% of the eventual cost of the plant.

A laboratory study can never totally simulate the conditions at full scale due to the highly variable nature of industrial effluents. On-site pilot trials are to be recommended in addition to laboratory treatability studies. Advantages include more representative data and better information on operating problems with industrial scale ancillary equipment. Factory staff can also be familiarised with the process.

5.4 Effluent treatment processes

5.4.1 On-site treatment

Simple on-site treatment can easily be justified, e.g.

(a) *Separation of solids waste*
  (1) Raw materials.
  (2) Processing waste.
  (3) Packaging.
  (4) Spoilt product.

The cost of disposal of these individual wastes can be much cheaper than a combined waste because they are easier to recycle or use as by-products, e.g. animal feeds. The negative value of refuse is −£35/tonne, 70% of this cost is for transport.

(b) *Consolidation*. Normally compaction and compression of packing materials reduces the volume and therefore the costs of disposal.

(c) *Effluent*. Drainage charges are typically between £1–3/tonne, the charge is based on strength and volume (Table 5.2), so some simple on-site treatment to remove solids and therefore strength is usually cost effective. Other schemes such as stream separation and finally biological treatment can also be justified by cost benefit analysis.

5.4.2 Pretreatment processes

Waste waters are a complex mixture of suspended and dissolved materials and complete treatment has to be a combination of various units (Fig. 5.1). Biological treatment, aerobic or anaerobic, is the major part of the treatment process removing most of the organic waste. Biological treatment can deal with both suspended and dissolved or colloidal materials but normally it is more cost effective to remove easily settleable solids physically. Most of the upstream processes are designed to protect the biological stage from shocks, blockage with solids or damage by inerts. The downstream processes are for biomass separation. Preliminary and primary treatments can be used independently of biological treatment and are usually a vital part

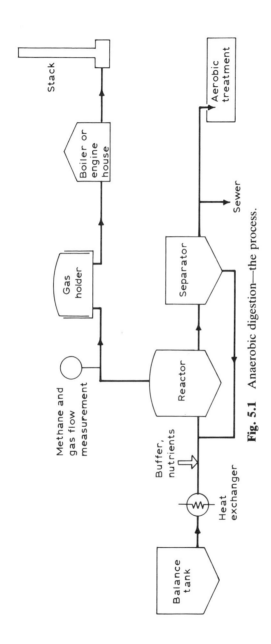

Fig. 5.1 Anaerobic digestion—the process.

of any treatment system. Additional or ancillary equipment is required for temperature control, pH control and gas handling. Anaerobic processes work best at 35–37°C and at a neutral pH.

(a) *Screening*. Screening and grit separation are intended to remove material which although not exerting a major polluting load on the biological treatment plant, could cause blockages and damage to equipment. Occasionally screening can remove considerable amounts of organic load but only that associated with large solids such as in the meat and vegetable industries. Screens are designed to strain out debris such as feathers, rags, string, plastic film, cardboard and other packaging material as well as any stray raw materials or product. Normally the minimum aperture size is 4–5 mm, smaller sizes block too quickly even for continuous cleaning. Some wastes require both coarse and fine screens because the fine screens block too quickly. Larger debris is removed with a screen spacing of 25–50 mm. It is essential to install some form of screens for even apparently solids-free wastes such as dairy effluents to prevent packaging materials damaging mechanical equipment. Food industry screenings can be recovered for use in animal feeds, otherwise screenings are incinerated or landfilled.

(b) *Grit removal*. Separate mud and soil removal may be necessary with certain wastes, to prevent wear on mechanical items such as pumps. It is usually essential for root crops such as sugar beet, potatoes, etc. The denser minerals are separated from organics in constant velocity channels, chambers or hydrocyclones. A velocity of 0·3 m/s is used to allow grit to settle whilst keeping the organic material in suspension. The deposited material is landfilled.

(c) *Flow balancing/two-stage treatment*. The flow and strength of an effluent often changes rapidly with time. There may be poorly biodegradable components, or extremes of pH and temperature. Biological treatment can deal with a wide range of materials including shocks, but it is normally cheaper to control these disturbances by flow and load balancing. All types of equipment work more efficiently when the flow and load are constant. Most Water Authorities also specify, in the consent, a maximum flow rate as well as the total amount of effluent to be discharged to sewer or water course. Balancing tanks smooth out rapid changes in the raw waste and are frequently a useful point for reactor control, e.g. pH and chemical or nutrient dosing. Balancing usually consists of a holding tank with mixing. Sludge and scum are common operational difficulties and some methods for their removal should be incorporated into the design. Forward flow is via a constant rate pump. An emergency overflow is also needed. The size should be based on the survey and waste audit prior to plant design. The undersizing of balancing tanks and associated equipment is a common problem in industrial effluent treatment. A holding capacity of between 75 and 100% of the total daily flow is usually required. This also ensures some time for hydrolysis of the waste prior to the anaerobic digester (see Chapters 2 and 4). Anaerobic treatment does not have the problems associated with oxygen depletion that aerobic treatment has and oversizing a balance tank is not detrimental. Anaerobic digestion is often enhanced by separation of hydrolysis and acidification prior to methanogenesis. Optimal conditions for acid hydrolysis are different from those for methanogenesis and separation is easy to achieve by the pH requirements and growth rates of the different groups of organisms.

(d) *Control.* The complex synergism between the microorganisms involved in the anaerobic breakdown of waste makes the process susceptible to shocks (see Chapter 2). Many studies have identified alkalinity as a key parameter which could provide early warning of impending problems. Wastes which generate alkalinity can tolerate acidic feeds. The anaerobic reactor pH can still remain within the acceptable range (7·0–8·0) even with a feed pH as low as 3·5 (Wheatley and Cassell, 1985). Previous experience or laboratory tests are essential to determine the characteristics of the waste. Simple wastes such as sugar confectionery, soft drinks, jam manufacture, etc. frequently give problems whereas complex wastes, e.g. brewery, milk or domestic do not. pH and alkalinity adjustment is usually necessary as a precaution. pH correction should be carried out between the balancing tank and the digester. The main difficulty with automatic control is probe fouling due to biological growth accumulation of solids and electrochemical ageing. Self cleaning systems are available, but continuous maintenance is still required. The most common alkalis used are sodium hydroxide and lime. pH control may be complemented by other sensors (see Section 5.6).

(e) *Settlement and flotation*. Settlement and flotation are cheaper than biological treatment and solids separation can reduce the size of the anaerobic reactor. The solid portion of an effluent is also the most difficult to treat biologically. Settlement is most useful for the removal of solids larger than 10 mm, flotation for smaller than 10 mm. Settlement or flotation of brewery and vegetable wastes for example could remove 30–50% of the total organic load and 60–70% of the suspended solids. Other wastes from soft drinks or sugar confectionery are unlikely to benefit. Settlement or flotation can prevent the formation of sludge deposits and scum elsewhere in the treatment system. Two types of anaerobic reactors are resistant to solids, i.e. the Contact Process and sludge blanket process (see Section 5.4). It may be cost effective to anaerobically treat the whole flow. Certain inorganic and organic trace nutrients may also be encapsulated in settled solids and lost to the anaerobic reactor during settlement. For fluidised bed reactors and anaerobic filters solids removal is very important to avoid clogging.

Settlement tanks are designed to produce quiescent conditions, by generating a uniform flow and by ensuring that settled particles are not resuspended. Typically retention times are 2–4 h with an upward flow velocity 2–3 m/h. Tanks are usually 2 m deep with a 1 in 5 to 1 in 10 sloping bottom. The feed into the tank is via a diffuser drum to avoid turbulence and short cutting. Slots may be cut into the base of the drum to avoid scouring of the bottom of the tank. A sludge scraping mechanism is essential, deep hopper bottom tanks common at very small plants are not satisfactory. Sludge accumulates on the tank walls and scum on the surface of the tank. A scraper ensures that the sludge is pushed and collected into a hopper at the centre of the tank. A scum baffle should also be fitted to the top of the scraping mechanism to collect floating material which occurs due to gassing and fats. A scum valve or slot is then required to remove this surface material. Sludge accumulating in the hopper of the tank is removed periodically during the day by hydrostatic head through a screw threaded bell valve.

Many industrial wastes contain fats and oils. Most are biodegradable, but those from the petrochemical industry are degraded at a slower rate than those from the food industry. The effective biological treatment of fats, oils and greases (FOG) relies on their solubility or dispersion to provide a uniform concentration accessible to the biomass. Unfortunately oils and grease tend to separate from the main effluent stream and solidify, blocking or blinding biological reactors and process instruments. It is usual to separate and recover fat. The warmer temperature of anaerobic treatment makes it less susceptible, compared to aerobic treatment, to fat accumulation. Typically 200 mg/litre of FOG is recommended as the maximum for biological treatment. Like settlement fat separators are designed to produce quiescent conditions to allow the density difference to bring fat to the surface. Wastes frequently contain detergents and surface active materials which emulsify the oils and therefore reduce the efficiency of separation. The efficiency is also reduced by warm temperatures. More sophisticated systems have been developed to overcome these problems. The most common alternatives to gravity separation are dissolved air assisted flotation and inclined plate separators. Fats and oils separated by these techniques can usually be recovered as by-products.

Dissolved air flotation is well established in waste water treatment, but there are no reports of its use prior to anaerobic treatment. There may be a risk of carry over of dissolved oxygen causing inhibition of the anaerobic reactor.

5.4.3 The anaerobic reactor

The type of design depends on the waste to be treated. Most of the organic pollutants in domestic sludge and animal slurry are as solids. The organic matter in industrial waste is in solution or colloidal suspension and therefore amenable to rapid treatment. The anaerobic bacteria grow slowly (see Chapter 2). If there is no special system for keeping and reusing the bacteria, then the minimum retention time in the reactor is limited by the microbial growth. The doubling time of the anaerobic methanogenic bacteria is 5 days and this is too long a retention time for reasonable commercial treatment of industrial effluent. New reactor designs have been developed for industrial wastes which hold back most of the organisms inside the reactor or which recycle the bacteria after separation. Bacterial retention is thus uncoupled from liquid retention and high bacterial concentrations can be obtained.

Bacteria are retained in the reactor by four basic methods (Fig. 5.2):
  (a) Physical separation of the biomass from the effluent by filtration or sedimentation, followed by recycle back to the reactor. This type of reactor is known as the Contact Stirred Tank Reactor (CSTR).

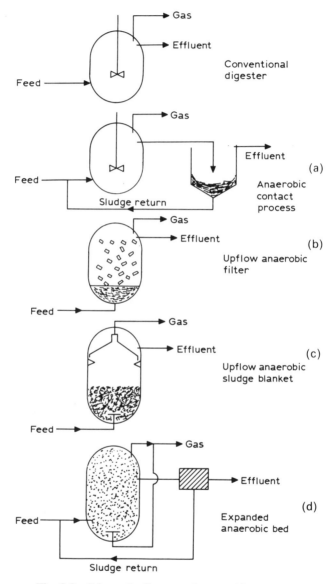

Fig. 5.2 Schematic diagram of anaerobic reactors.

(b) Retention and attachment of the bacteria by an internal packing to reduce upflow velocity; anaerobic filters.
(c) Natural bacterial flocculation assisted by low upflow velocities, known as UASB or Upflow Anaerobic Sludge Blanket.
(d) Attachment of the bacteria to a small support particle and fluidisation to produce mixing (fluidised or expanded bed).

There are also a variety of other reactor types; either bulk volume low rate systems such as lagoons, and large stirred tanks as well as some hybrids. These are not considered in detail in this chapter. Low rate very large lagoons are very cheap to build, but they are really only suitable in warm climates with large areas of low cost land.

(a) *The contact stirred tank reactor.* This was the first type of anaerobic effluent treatment system to be used. The idea was developed from the anaerobic digestion of municipal sewage sludge (Department of Scientific and Industrial Research, 1960). The principle was modelled on the activated sludge process.

The bacteria are physically separated from the effluent by settlement or filtration and recycled back into the reactor. Unfortunately anaerobic bacteria do not precipitate easily. The warmth of the effluent (relative to ambient) and the methane gas being given off make the solids buoyant. Settlement has to be assisted by degassing, cooling, filtration or inclined plates.

The early designs (Morgan, 1980) used cooling by heat exchanging with the incoming waste. This reduced thermal mixing in the settlement tank and inhibited gassing by slowing down methanogenic activity. Problems were still encountered and the present designs include much larger settlement tanks (Anon, 1987a). Large inerts in the feed have also been found to act as high density nuclei or carriers for the biomass and thus wastes which contain a high solids content would be suited to this process. Upward flow velocities in the settlement tank should be less than 0·5 m/h. Scandinavian designs use a preflocculation/degassing stage before quiescent settlement. This is carried out in a small tank with a low speed stirrer to create gentle mixing (Frostell, 1981). The treated waste passes on to settlement augmented by inclined plates. The process has been successfully applied to treat the waste from sugar beet processing (Shore *et al.*, 1984).

Another approach to overcome the problem of settlement has been to use membrane filtration to concentrate biological solids (Butcher,

1988). This approach was successful for the treatment of digester solids and has justified the high running costs by the quality of effluent produced. The life of the membrane has exceeded expectations and ran for six years before replacement.

MIXING. Good mixing is also necessary in a CSTR to ensure uniform substrate and biomass concentration. A national survey of mixed anaerobic digesters treating municipal sludges in 1969 (Swanwick *et al.*, 1969) revealed that it was common to find half the reactor ineffective because of poor mixing. Most of the problems were traced to blockage, fouling and corrosion of mechanical mixers. Traditionally mixing has been by paddles in a draught tube. Much of the mechanism of the paddle type is immersed in the digester and maintenance is very difficult.

Another common method of mixing is by pumped recirculation of the digester contents. Often, however, the less dense sludge is preferentially circulated leading to stratification and leaving much of the digester unmixed. The older designs of digester were too large for easy mixing. Modern digesters are half the size and mixing is much easier. Gas lift mixing is popular in these types of digester. Biogas is compressed and reintroduced into the base of the digester either via an unconfined ring, or into the bottom of a draw tube.

APPLICATIONS. The CSTR is more complex than either the UASB or anaerobic filter but the absence of any internal fittings makes it suitable for the treatment of wastes with a high solids content. The two critical design features, settlement and mixing, mean that the typical loadings are modest compared with the alternative reactor types (Table 5.5).

The relatively low loadings make the reactor best suited to difficult wastes where land is readily available. At typical retention times, 2–5 days, it does not require close supervision and is quite resilient. The CSTR system is still cheaper than aerobic treatment because it generates little surplus sludge and has low energy requirements. The

Table 5.5 Typical operating conditions for a CSTR

Load	0·5–2·5 kg COD m^3/day
Retention	1–5 days
COD removal	80–90%
Solids concentration	5–10 000 mg/litre

CSTR is a popular type of anaerobic treatment. About 30% of industrial anaerobic treatment plants in Europe are of this type. Much of its popularity is based on its similarity to conventional municipal sludge digestion. About half of these reactors do not recycle biomass at all with an average retention time of about 7 days. All the applications are in the agrofood industry. The most common waste is from the distillery industry (15 full-scale plants, 10 in Italy) (Demuynck et al., 1984). Treatment rates with this waste are very low, retention times about 10 days. They vary in size from 300 to 12 000 m^3 reactor volume. There are also about 10 full-scale anaerobic contact plants treating sugar processing waste water, mostly in Belgium and France. There are several full-scale contact digesters in the UK. One treats a fermentation waste and has encountered significant sulphate toxicity difficulties. It is unrepresentative. Some design details of a reactor to treat brewery waste have been published (Anon., 1987b). Several papers have been published on one of the earliest to be built to treat starch processing waste waters (Morgan, 1980; Butler, 1984; Butcher, 1988). This reactor treats a waste with a COD strength of 2000 mg/litre at a load of about 2·5 kg COD m^3/day and a retention time of 24 h. COD removal is 70–80%. Difficulties were experienced with maintaining a sufficient biomass concentration in the digester by quiescent settlement. Eventually the sedimentation step was replaced by a membrane filtration unit which separates the biomass from five 200 m^3 digesters on rotation. The plant has worked well ever since. Data have also been published on a second CSTR at the British Sugar Corporation, Bury St Edmunds (Shore et al., 1984; Smith, 1984). The plant treats half the total waste from a sugar beet processing factory during the six month season. The organic load is 2·5 kg COD m^3/day at a retention time of 2·5 days. Examples of results for the 82/83 and 83/84 season are presented in a paper by Shore et al. (1984) which shows an average COD removal of 86%. The digester is 6600 m^3 and the flocculator 50 m^3. No difficulties were reported on keeping the biomass in the reactor. The concentration increased steadily during the season to reach about 5 g/litre. Details of the economics are given in Section 5.7. This plant has been a success and other digesters have now been built.

THE ANAEROBIC FILTER. The difficulties associated with settling and recycling biomass led to an alternative approach based on aerobic percolating filters. Bacteria were immobilised on a support medium.

The reactor is plug flow either up or down and includes a packing which must provide a large surface area for microbial attachment while including sufficient voidage to prevent blockage and channelling.

PACKING MATERIALS. In the first anaerobic filters, the support media were natural materials such as stone and gravels. These had a very low voidage, <50%, and blocked with biomass and solids very quickly (Young and McCarty, 1969). Plastic rashig rings or similar were tried. This prolonged the time intervals between blockage but did not overcome the problem totally. Downflow filters were then tried which ran at high liquid velocity. The high irrigation rates were achieved by introducing recycle. Recycle improved the wetting of the surface provided which is more difficult in a wetted rather than a submerged system. Blockage and channelling were also evident in this system particularly when the waste contained fat, insoluble organics or a high concentration of solid material (Environment Canada, 1986). Present design practice anticipates clogging and includes facilities for washing the media by injecting recycled effluent and gas to dislodge accumulated biomass and solids (Camilleri, 1987; Ehlinger et al., 1987). Once the pack has been washed then performance is recovered. Random plastic support media were also found to be easier to fluidise and mobilise by recycle. The proportion of the reactor containing packing was also reduced and is now typically 50–70% (Oleszkiewicz et al., 1986; Oleszkiewicz and Thadari, 1988).

Table 5.6 gives some characteristics of typical packing materials. The natural stone media should not be used, the voidage available for the accumulation of solids is insufficient. Voidage should be at least 90%. Plastic or ceramic rings are the most common materials. The type of plastic does not affect reactor performance. Most surfaces may be colonised by bacteria. Nylon, polypropylene and polyvinyl chloride have all been used. Attachment to hydrophilic surfaces is very rapid. If

Table 5.6 Some properties of typical media

	Wt (kg/m^3)	Specific surface (m^2/m^3)	Voidage (%)
PVC Flocor random 50 mm	92	300	94
PVC Flocor E ordered	90	100	95
Cloisonyle	80	200	94
Furnace Slag	975	108	55

the surface is hydrophobic, like the plastics, then the extra-cellular polymeric secretions of the bacteria act as a bridge between the surface of the cell and the plastic. The attachment process occurs in two stages: (a) an initial electrostatic attraction to the surface with surface roughness providing shelter against the liquid shear forces, followed by (b) permanent binding by the extra-cellular secretions. Removal of the bacteria after this type of attachment requires strong treatment. Close examination of the solids held in an anaerobic filter reveals three different types:

(1) fine loose solids 1–10 μm in size held in equilibrium by the upward fluid velocity;
(2) flocculated granulated biomass and solids in the interstices of the media. This material constitutes the greatest amount of the active biomass and is the source of the clogging problems;
(3) bacteria firmly bound and closely following the contours of the support media.

The specific surface available in the filter is not as important as in aerobic biofiltration because most of the active culture is held as flocculated growth rather than attached to a surface. The simplest and cheapest support media may therefore be used. The media must be mechanically stable and able to tolerate some long term compression.

There is no published information on the correct size of the media. The larger random media (i.e. >100 mm) would resist clogging better than the smaller 25–50 mm medium but may be more difficult to fluidise in the washing process. Most of the solids accumulate at the base of the reactor so it may be possible to grade the support media with the smallest at the top.

A variety of materials have been used for downflow filters including ceramic blocks, bricks and polyester sheet. They are cheaper than the proprietary supports.

HYDRODYNAMICS AND MIXING. In a plug flow reactor much of the effective treatment efficiency will take place in a highly active zone in the region where the waste is introduced into the reactor. Thus a substrate concentration and activity gradient will be established through the height of the reactor (Dahab and Young, 1982; Cheung et al., 1986). This leaves some reserve capacity for shock loads and other upsets. Plug flow biofilm processes have been shown to be more stable and reliable than CSTR reactors. There may be other factors such as close physical contact between species for symbiosis (see Chapter 2) not available in completely mixed systems. By the same simple model, however, it must also be true that completely mixed systems are more efficient in the use of reactor volume. CSTR reactors normally achieve a better percentage COD removal. A crucial feature of a plug flow system, therefore, is the introduction of the feed. Maldistribution or dead spaces due to clogging will lead to short circuiting, reduce the retention time and consequently the performance of the reactor.

Typically distributors are like 'cartwheels' with a number of arms depending on the diameter of the vessel. The distribution holes point downwards into the sludge bed with the diameter of the holes, or number of holes, profiled away from the pumped input. The recycle is introduced through a larger totally separate, but similar system slightly above the feed distributor. A minimum of up to five times the forward feed rate should be provided for recycle. The distributor arms should also protrude through the reactor walls to allow some possibility of rodding out. The treated effluent and the recycle are collected in 'V' notch troughs at the top of the reactor. In the case of a downflow reactor the system is simply reversed.

The amount of dead space or clogging can be measured by standard techniques (Levenspiel, 1972; Wheatley *et al.*, 1988) and used to determine washing frequencies. Highly active filters with gassing and recycle are nearly completely mixed. This will overcome deficiencies in distribution but not those due to clogging.

APPLICATIONS. The simplicity and robustness of the anaerobic filter, i.e. its resilience to shocks, load, pH, substrate, character, etc. make it ideally suited to the treatment of soluble industrial waste. Loading rates are much higher than in the CSTR (Table 5.7). The anaerobic filter is not as popular as the CSTR or UASB reactors. The major worry is excessive biomass accumulation and clogging of the packed bed. Fifteen percent of the industrial anaerobic treatment plants in

Table 5.7 Operating conditions of an anaerobic filter

Load	2–10 kg COD m^3/day
Retention time	10–50 h
COD removal	70–80%
Critical solids concentration in the feed	1 000 mg/litre

Europe are of this type. Three full-scale anaerobic filters are used for the treatment of chemical wastes from the textile industry (Witt et al., 1979). They treat a mixture of alcohols and esters.
There were no reports of blockage although the solids balance is measured and provision has been included for periodically cleaning the reactor. The remaining filters are in the agrofood industry. There are at least ten anaerobic filters in France treating distillery and sugar wastes (Camilleri, 1987). Results have also been published from a very large downflow filter (1200 m^3) treating distillery waste at Bacardi (Szendry, 1983, 1986). The retention time is 10–15 h, COD removal is 80% at a load between 3–5 kg COD m^3/day. This filter was built in 1982, there are no reports of blockage. A downflow filter with natural media (Environment Canada, 1986) treating dairy waste did block with fat after two years of operation. The choice of the anaerobic filter must be governed by the waste. Wastes with a high solids content or fat content will cause problems. Some sugar-bearing wastes also generate mucilaginous growths (Ehlinger et al., 1987). Provision should be included in the design of all filters for media cleaning by gas or effluent recycle. There are two anaerobic filters in the UK. One built to treat distillery waste has never achieved any significant COD removal because of sulphate toxicity. The second, a demonstration scheme, to treat confectionery waste has encountered significant shock load problems (Wheatley et al., 1984).

(b) *The Upflow Anaerobic Sludge Blanket Reactor (UASB)*. Problems due to clogging of the filter media in anaerobic filters, particularly the early mineral packings, resulted in investigations into methods of flocculating the bacteria within the reactor. A high density granular sludge is needed which may then be retained within the reactor despite the gassing and upflow velocity of the waste. UASB reactors require a gas/liquid/solid disengagement zone. This is a system of baffles at the top of the reactor (Fig. 5.3).

GRANULATION. Sludge granulation is a complex and not fully understood process. Comprehensive and detailed studies have been made in a number of countries but principally in Holland where the UASB process was devised (Lettinga et al., 1987). To maintain process efficiency settling velocities of 10 m/h are required. Measured velocities up to 90 m/h have been reported (Lettinga and Hulshoff-Pol, 1986). Granulation is a natural process and is due to a combination of microbial morphology, nature of the substrate and accumulation of

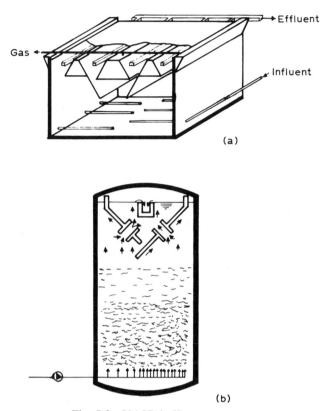

Fig. 5.3 UASB baffle arrangement.

inorganic salts. Formation of rapidly settling granules from an ordinary inoculum may take 50 days (Alibhai and Forster, 1986). Key elements in the feed substrate for successful formation are calcium, phosphorus, aluminium and silicon. These are readily available in root vegetable processing waters which contain some clay and soil. A large population of filamentous organisms, *Methanothrix* sp. are also essential and may be associated with a high concentration of soluble components in the substrate, mainly acetic acid. A third factor is the generation of bacterial polymers. The type of waste water to be treated has an important role in the formation of granules. A satisfactory sludge blanket was formed from yeast processing, sugar beet and potato wastes but not with distillery, corn starch, abattoir or dairy effluents (Stronach *et al.*, 1986).

Fig. 5.4 Photographs of the UASB process. (a) A plant in Holland treating potato waste showing aspect ratio; (b) a plant in Italy showing the system for carefully controlling flow distribution.

To reduce the time required for acclimatisation and adaption it is now usual to start with a large inoculum, of an already granulated sludge. Recommended amounts are 10 kg volatile suspended solids (VSS)/m^3 of reactor volume.

HYDRODYNAMICS AND MIXING. Baffles inside the UASB reactor are used to promote gas/solid separation to retain the granules. Two examples are shown in Fig. 5.3. Solids collecting in the cones should settle back into the digester after releasing the gas. Normally these baffles can be made of reinforced plastic. Distribution and upward flow velocity are also important for granule retention. The degree of mixing in the sludge blanket is a function of the gas production and the upward flow velocity of the influent. Reactors have a low aspect ratio with a large plan area and large number of small distribution points (Fig. 5.4) (Lettinga *et al.*, 1983). The effect of gas production is more complex. Higher than optimum gas production will cause gulping and open paths through the blanket, lower than optimum gas production will not encourage enough mixing within the blanket to avoid short circuiting through the dense sludge granules. This sets limits on the required reactor height and number of distribution points. It has been shown (Bolle *et al.*, 1986) that at typical gas and liquid velocities (1–1·5 m/h) then a sludge blanket height of 3·5–4·0 m in a 6·7 m reactor is optimal.

APPLICATIONS. The UASB has been extensively applied and there is much published information. Typical operating conditions are shown in Table 5.8.

The UASB is the most popular type of anaerobic industrial waste treatment (30% of the plants built). It has been very successfully applied in the potato industry and sugar beet industry. There are about 40 plants in Holland alone and sizes are between 100 and 1500 m^3. Problems have been encountered with applications to dairy and slaughterhouse wastes. Losses of fine solids and periodic biomass

Table 5.8 Typical operating conditions of a UASB

Load	2–15 kg COD m^3/day
Retention time	8·50 h
COD removal	70–90%

wash out have been encountered (Sixt and Sahm, 1987). There are five in the UK (Table 5.1), one is for the treatment of a paper mill effluent commissioned in 1986, no monitoring data have been published (Anon., 1987b). The second treats dairy waste and is an energy demonstration scheme. The plant has encountered some problems with fine solids and sludge retention. The main operating problems of UASB are start up and maintaining the flocculated granules.

(c) *The fluidised bed.* The difficulties of biomass separation in completely mixed reactors and the loss of granulation or blockage in plug flow reactors can be overcome by a combination of the two types. This is the basis of the fluidised bed.

MIXING AND HYDRODYNAMICS. The support particles are expanded and mixed by gas and effluent recycle. Sufficient recycle to cause some lateral mixing of the particles and avoid channelling is necessary. Bed expansion is typically 20–25% of the reactor volume. Unlike other completely mixed systems the design of the distribution system is critical since this is responsible for the fluidisation of the bed. Most designs incorporate perforated distributor plates or inverted cones. The minimum recycle ratio is governed by the density of the particles. In the case of natural particles–sand, silica or carbon, then the upflow velocity will be set by the amount needed to fluidise the uncoated particles. In the case of plastic particles then this velocity will be much lower. A velocity ratio of 1·2–1·5, the minimum for fluidisation of the uncoated particles is necessary (Shieh and Mulcahy, 1983). Some continuous adjustment of the recycle pump to compensate for changes in biomass density will always be necessary.

A second distinct zone is usually incorporated within the fluidised bed for clarification. This can be an increase in the diameter of the top 20–25% of vessel. Foaming can be a problem with certain protein-bearing wastes and if the fluidisation energy required is high. This may be counteracted by an antifoam agent and or an upper grid to avoid particle losses. Support particles are of two types—natural materials such as sand, anthracite, gravel, silicon or carbon, and various types of plastic. The plastic materials can be fabricated into porous structures and have the advantage of a lower density. Much of the effectiveness of a fluidised bed compared to an anaerobic filter can be attributed to the smaller carrier particles which may be used without the risk of clogging. Diffusion may become limiting in porous particles. The

Table 5.9 Operating conditions for a fluidised bed

Load	2–50 kg COD m^3/day
Retention time	0·5–24 h
COD removal	70–80%

ideal characteristics of fluidised bed particles have been investigated (Walker and Austin, 1981; Switzenbaum, 1983). Very small particles give the best performance, i.e. 0·5–0·2 mm but may be difficult to retain within the reactor.

APPLICATIONS. The greatest risk with the fluidised bed is loss of the active biomass from the reactor by sudden changes in density or gassing. Fluidised beds have therefore acquired the reputation for being difficult to operate. Biomass concentrations in fluidised beds are higher than the alternative reactor types (10–40 g volatile solids/litre) and potential performance is very high (Table 5.9). Application of the technology may depend on the perfection of new control systems.

Fluidised beds are relatively new and few plants have been built so far. There are five in Europe, details have been published on two treating yeast processing wastes (Fig. 5.5). There are none in the UK. Much of the work reported is at laboratory or pilot-scale and most of the general types of waste have now been tested. Jeris (1983) and Heijnen et al. (1985) report on studies on whey, starch, soft drinks and chemical wastes, Rudd et al. (1985) on meat wastes and Stronach et al. (1986) on pharmaceutical wastes.

5.4.4 Materials of construction

Traditionally anaerobic digesters have been large and built of reinforced concrete. These are too expensive for most industrial waste and standard prefabricated tanks are now commonly installed. Typically they are between 200–500 m^3 in size, larger sizes are obtained by using several tanks. This modular approach also gives more reliability. Mild steel bolt together tanks can be used, the sulphides formed by the low redox potential within the reactor, give protection against corrosion. There is, however, always the possibility of excess acidity within the reactor and it is usual to coat steel with enamel. Concrete is treated

Fig. 5.5 A two-stage fluidised bed treating yeast processing effluent. Photograph: Gist Brocades.

with epoxy paint. Digesters are insulated with waterproof synthetic materials or rock wool with weatherproof cladding.

5.5 Ancillary equipment

5.5.1 *Heating*

Anaerobic digestion works best between 35–40°C (mesophilic range). Many industrial wastes particularly from food processing, are warm but not reliably in the correct temperature range. Some form of heating is required. Virtually all types of heater have been used, but traditionally internal coils are the most common. They have the advantage of simplicity but the coils encourage fouling and slowly lose their efficiency due to 'baked-on' biological solids. External heat exchangers are preferred for easy maintenance. Plate and frame exchangers are generally the most efficient, but can really only be used for solids-free wastes or in conjunction with automatic cleaning. Shell and tube heaters are therefore the most common.

5.5.2 *Gas storage*

Storage of the gas formed during digestion is expensive but necessary to even out the biological fluctuations in production. For use in boilers and engines only a few cm of water gauge (150 mm) are necessary to run the equipment. The aim should be to plan to use the gas as it is produced and avoid storage. Typically 5–10 h storage are provided. A number of digester designs incorporate larger headspaces at the top of the reactor as a way of reducing the volume of gas storage. Floating steel gas-holders are very durable and the most common. Prefabricated reinforced plastic holders are available and are more resistant to corrosion. The maximum size is about 50 m^3 because they cannot be made strong enough at larger sizes. Variable pressure butyl gas balloons are also available. They are cheap and convenient to install but are prone to damage.

5.5.3 *Biogas utilisation*

The cheapest and most effective use of the gas produced is direct combustion and most boilers are normally suitable. If the gas is particularly dirty (i.e. containing water and sulphides) then more care has to be taken over the materials of boiler construction, using for example ceramic burners. The second most common use of the gas is power generation in high compression spark ignition engines (range 10–100 kW). The heat from the cooling water may be recovered to warm the digester.

5.5.4 Settlement

Post-anaerobic settlement or separation is essential for CSTR reactors. Sedimentation can also remove or recover substantial amounts of biomass or solids after the other types of reactors. Decisions on installing post-reactor settlement for these reactors depends on local conditions such as water authority consent limits, type of waste, and type of aerobic treatment.

5.5.5 Aerobic treatment

Anaerobic treatment processes operate at a low redox potential (-350 mV) and this means that organic nitrogen and sulphur compounds are reduced by the process to ammonia, amines and various sulphides. The smell, oxygen demand of these compounds and their toxicity means that anaerobically-treated effluents are unsuitable for discharge to inland water courses. Reaeration is required. There are two convenient methods of reaeration: aerobic polishing treatment or discharge to sewer for aerobic treatment in combination with domestic sewage. The local water authorities are likely to impose concentration limits for sulphides, amines and methane for discharge to sewer. In some parts of Europe, for example, France, and some areas of the UK, nitrogen compounds attract an extra treatment charge.

5.6 The control of digestion

5.6.1 Safety

Methane at between 5–15% in air is explosive, and there are safety recommendations concerning its use and handling. Anaerobic plants must be fenced off from other areas and have restricted access. It should be sited as far from other buildings as possible and all electrical equipment should be at least 3 m from the digester and gas-holder (BABA, 1982). A methane safety meter should be provided for checking concentrations in the buildings for instrumentation and pumps. The buildings should be well ventilated. Flame traps must also be provided in the pipelines from the digester and gas-holder. Pressure relief valves and gauges to indicate blockages must also be provided for the digester, gas-holder and major pipelines.

The second hazard is from hydrogen sulphide in the gas. Hydrogen sulphide is toxic at 10 ppm and the standard practices need to be followed concerning confined or enclosed spaces. A portable gas alarm for sulphide will also need to be provided.

5.6.2 Instrumentation

Methanogenesis from organic materials relies on a diverse group of microorganisms (described in Chapter 2). It is easy to disturb the balance between these different organisms. The most common difficulty is a sudden increase in the strength of the waste which leads to a rapid acceleration of acidogenesis, but a much slower response from methanogenesis. This can cause excess acid and an inhibitory drop in pH. Few industrial effluents are sufficiently buffered or contain enough additional nutrient to give reliable performance from biological treatment at a high rate. This means anaerobic digestion is vulnerable to sudden changes in the characteristics of the waste. Shock loads are typical of industrial waste and these potentially inhibitory changes in waste character must be detected and controlled. The strength of the waste may be measured retrospectively by chemical analysis, i.e. BOD, total organic carbon (TOC) and COD, but there is no convenient method of rapidly determining the organic strength of effluents. There is much research in the development of a simple on-line COD probe.

The alternative approach is to utilise a combination of existing sensors linked together via an 'expert system' to give an integrated, indirect prediction of waste strength. This feed forward parameter can be used to control forward feed rate, addition of chemicals, an increase in recycle rate, or diversion to a holding tank. This controlled response may also be modified by a feedback system which gives an indication of the reactor's ability to cope with an incoming shock.

(a) *Choice of sensors.* Modern sensors need to provide on-line information and be easy to operate. The main problem associated with on-line monitoring is probe reliability. Difficulties are encountered due to biological growth, accumulation of other solids and electrochemical ageing. Electrode calibration and measurement of the degree of fouling can be incorporated into the software. Experience has shown that probes in the reactor or inserted into pipework are rapidly contaminated and cannot be restored. The probes should be in open tanks wherever possible. The feed forward probes can be installed in a separate weir box used for flow measurement.

GAS ANALYSIS. Gas analysis is the most useful on-line performance indicator. It is non-invasive. There should be a roughly constant production of methane from the same amount of organic matter metabolised, i.e. about 0.5 m^3 methane/kg of (volatile solids) VS or 0.35 m^3 CH_4/kg COD removed. An average mass balance can be established based on typical removal of VS and feed rate. Methane production is sensitive to normal variations in operation such as periodic feeding but this can be overcome with continuous monitoring and adjustment based on a moving average. The methane content of the gas is typically 50–75%. This can be measured by infra-red spectroscopy, thermal conductivity or calorimetry. The normal steady state gas production and methane content can be established and a deviation used as an early indicator of digester problems. Gas production is affected by temperature and is likely to vary a little with ambient temperature and feeding.

TURBIDITY. A turbidity meter will give information on the suspended solids in the waste. In many wastes, i.e. dairy and fermentation effluents, solids can represent 50% of the organic substrate. A turbidity meter can be used in conjunction with other probes to assess the strength of the incoming waste.

AMMONIA. The ammonia probe is also established in the water industry. It is used principally as a water quality monitor. Recent work (Naghdy and Helliwell, 1987) has established a relationship between strength and ammonia in domestic sewage. It is likely that it may also be used with industrial wastes containing a significant amount of organic nitrogen.

CONDUCTIVITY. Conductivity like ammonia is common in water quality monitors for determining the concentration of soluble salts in water. It is not well established for monitoring waste water. Tests would need to be conducted with individual wastes to determine whether there was a relationship between soluble COD or BOD and conductivity.

TEMPERATURE, pH AND FLOW. Temperature, pH and flow are an important part of determining overall process stability. Four measurements of flow are necessary:

(i) crude flow to balancing
(ii) forward feed rate to the reactor
(iii) recycle rate
(iv) gas flow rate

pH would need to be measured before and after adjustment as would temperature.

(b) *System and software.* Data can be collected and processed by a microcomputer via a suitable interface. The same system can then be used for controlling feed rate, recycle and additions via a relay system. A simple model will need to be developed from published data and experience to establish the limits of the organic loading. This process should be no more complex than using a programmable logic controller (PLC).

5.6.3 *Laboratory tests*

(a) *Alkalinity.* The pH of the reactor and effluent are rarely sensitive enough for control purposes. A disturbance to the balance between methanogenesis and acidogenesis is accompanied by an accumulation of carboxylic acids but these are usually neutralised by buffer salts present in the fermentation. Once all the buffer has been exhausted then the undissociated acids accumulate and depress pH. Variable amounts of buffer, in the form of bicarbonate, are generated in the fermentation process from CO_2. To maintain permanent stability then 1000 mg/litre bicarbonate alkalinity are required (Parkin and Owen, 1986).

(b) *Biological tests.* Measurements of biological activity are limited because the tests are complex and time consuming. The tests are useful to identify toxins or nutrient deficiencies when the response time is not critical. There are three basic tests:

GENERAL ACTIVITY. Anaerobic digestion should produce a fixed amount of gas per weight of COD removed or weight of biomass (as VSS). These performance ratios can be established in small laboratory activity tests for unknown wastes and compared with standard substrates such as acetate and yeast extract (Speece, 1988).

ATP. ATP is one of the basic electron transfer or energy transfer units in biological systems. It may be extracted physically or chemically and then measured using bioluminescence. It does not differentiate between organisms but can be used as a general indicator of activity.

F_{420}. F_{420} is a specific coenzyme used in methanogenesis. It may be measured in its oxidised form by fluorescence and is a good indicator of the numbers and activity of methanogens (Stronach et al., 1986).

5.7 Economics

The costs of the anaerobic treatment of strong industrial wastes are cheaper than the alternatives. Table 5.10 is an estimate of the running costs (power and sludge disposal for a vegetable and yeast-processing treatment plant).

The capital costs are shown in Table 5.11. The value of the gas produced is not included and this is estimated to be worth about a further £12K per annum. The most important savings are in aeration and sludge disposal costs. Often, because of the efficiency of aerobic metabolism 1 kg of substrate as BOD can be converted into 0·9 kg of dry solid. Unless there is a very cheap method of sludge disposal then

Table 5.10 A comparison of the cost of aerobic and anaerobic treatment

£ per kg of COD removed	Activated sludge	High rate filter (based on Flocor)	Anaerobic treatment (anaerobic filter)
Power	0·55	0·02	0·01
Sludge disposal (direct to local land)	0·20	0·05	0·002

Table 5.11 Summary of costs of treatment options[a]

Discharge to sewer without treatment: £311 400

Treatment type	New biotower, new foundations Option 1 (£)	New biotower, existing foundations Option 2 (£)	New anaerobic filter Option 3 (£)
Capital	655 200	523 500	630 000
Plant running costs and effluent disposal costs	224 612	224 612	143 026
Annual saving	86 780	86 780	168 297
Simple payback in years	$\frac{655\,200}{86\,780}$ 7·55	$\frac{523\,500}{86\,780}$ 6·03	$\frac{630\,000}{168\,297}$ 3·74
Payback, in years, by tax computation for cash flow method	7·67	6·25	4·08

[a] Taken from Smith et al. (1988).
NB The value of surplus gas—(Option 3) estimated value £12 000 pa in terms of coal equivalent has been ignored.

aerobic treatment is much more expensive than anaerobic treatment. Very low organic loads are required (i.e. <0·5 kg BOD m^3/day) for aerobic treatment plant design if surplus biomass production is to be similar to that from anaerobic treatment.

There are, however, still relatively few anaerobic industrial waste treatment plants, particularly in the UK. The process is still novel and includes a significant technical risk. The examples shown in Tables 5.10 and 5.11 are based on using anaerobic treatment as pretreatment prior to discharging the partially treated waste to sewer. The effluent after anaerobic treatment is in a reduced state and contains methane, sulphides, ammonia and amines. They are not suitable for direct discharge to a water course except the sea or a large estuary. Two UK water authorities have formally objected to anaerobic pretreatment because of a possible risk of the release of methane and sulphides in sewers. One has also requested guarantees, in the event of total failure of the anaerobic plant, that untreated effluent will not be discharged to the sewer to cause a sudden shock load. Anaerobic treatment should however still prove to be cost effective when it is intended to be the first stage and followed by aerobic treatment for discharge to the sewer or local water course. Simple post-reactor treatments such as flash aeration or chemical oxidation may also be suitable. The labour costs are similar, the data in Table 5.11 are based on a 24 h desludging cycle. The skill and training required to operate an anaerobic plant would be significantly greater than that for an equivalent aerobic fixed film biotower. Activated sludge plants are difficult to operate unless operated at low loadings.

The cost of the actual reactor is less than half the total cost of an anaerobic system (Table 5.12). The proportion of the costs is similar to all forms of effluent treatment.

Table 5.12 shows the published figures for the Bacardi plant. Costs in the UK vary between £200–800/m^3 of reactor inclusive of ancillaries but exclusive of civils and boiler changes. The range depends on the size and reputation of the company offering the technology and local site conditions, i.e. foundations, services, land availability, etc.

Some costs have also been published on the British Sugar plant at Bury St Edmunds (Shore et al., 1984; Smith, 1984). The capital cost was £100/m^3 in 1982 inclusive of ancillaries as above. It treats

Table 5.12 Capital and running costs of a 1 600 m^3 anaerobic filter ($—1981)

	Millions $
(1) Anaerobic filter and balance tank	3·12
(2) Civil works	2·59
(3) Ancillary equipment, gas handling, services, etc.	0·66
(4) Boiler modifications	0·16
(5) Electrical and instrumentation	0·40
(6) Design	1·2
Total including some extras	8·34
Operating Costs 1982	
	$ 000s
(1) Manpower	Not given
(2) Sludge disposal presumed discharged to sea	Not given
(3) Power	100
(4) Chemicals	25
(5) Maintenance	Not given
Total	275

2400 m³/day with a COD strength of 7000 mg/litre and generates about £200 worth of gas a day.

5.8 Treatability problems

5.8.1 Types of inhibition

In principle, biological waste purification is the most cost effective treatment for organic materials in solution or colloidal suspension. Purification depends on two mechanisms:

(i) Microbial metabolism of soluble organic substrate.
(ii) Adsorption of particulates onto biological floc.

There are some common problems with industrial waste water:

(i) Lack of additional nutrients.
(ii) Variability and waste shock loads.
(iii) Persistent toxins.

Many of these difficulties can be predicted and overcome by test work on the effluent before design (see Section 5.2). Microorganisms have a very rapid growth rate and in many cases it is possible to acclimatise and induce microbial metabolism of synthetic materials by selective pressure.

5.8.2 Nutrient requirements

For optimal growth microorganisms require a wide range of nutrients as well as an accessible source of carbon. There are three principal additional nutrients, nitrogen, phophorus and sulphur, which make up about 12, 2 and 1% of dry bacterial biomass. Anaerobic metabolism is not efficient and the growth rate is about 10% that of aerobic organisms. Nutrient requirements are 700:5:1 BOD:nitrogen: phosphorus (Sahm, 1984). This compares with 100:5:1 for an aerobic process. Few industrial waste waters contain sufficient nitrogen for optimum anaerobic growth. Even food industry wastes are not normally nutritionally balanced, and with the exception of fermentation wastes (i.e. yeast-processing, distilling, brewery and pharmaceutical wastes) they require nitrogen supplementation. Confectionery, starch and fruit processing wastes usually contain very little nitrogen. Nitrogen is, therefore, the key element and usually has to be added, phosphorus is usually present in excess.

There is also a second beneficial effect of nitrogenous compounds. During anaerobic fermentation ammonia is released from organic nitrogen compounds and this contributes to buffering by forming the ammonium ion (NH_4^+) (see Section 5.6).

Wastes which are rich in nitrogen can be treated at higher loads, i.e. 10–20 kg COD m^3/day if the nitrogen if 500:5 or better, but if deficient in nitrogen then the loadings must be more modest, i.e. 2–4 kg COD m^3/day at 1000:5. Unfortunately it normally proves very expensive to modify the C:N ratio.

In some cases more than one waste can be mixed together to provide nutrients. Gerletti (1984) describes a digester treating winery waste, and olive oil wastes which were mixed with urban sludge and animal residues in a large anaerobic digester.

(a) *Trace nutrients.* In recent years it has become apparent that the methanogens are very different from the other more common bacteria in the environment (see Chapter 2). Their cell walls are different, they have distinct genetics and they require unique trace nutrients. The unusual biochemistry of the methanogens means that they require slightly different cofactors and trace nutrients for optimum growth. Four metals have been shown to be required in quite high concentrations (Sahm, 1984):

Fe	2 nM
Co	10 nM
Ni	100 nM
Mo	10 nM

Manganese, zinc, copper, selenium and tungsten are also necessary together with the more usual sodium, potassium, calcium and sulphur.

Most of the transition metals are unfortunately insoluble at the very low redox conditions of the anaerobic fermenter. Organic sulphur and sulphate materials in the substrate are reduced to sulphide and this leads to precipitation of these metals as sulphides. Industrial waste waters may not contain adequate available concentrations of metals, and supplementation can improve performance during start up and shock loads (Speece, 1983).

5.8.3 Toxic materials

(a) *Metals.* Traditionally there has been a great deal of interest on the inhibition of anaerobic digestion by metals (Mosey, 1976; Parkin and Owen, 1986). This was because metals accumulate in the sludge/solids fraction in sewage treatment. Toxicity depends on the concentration of the soluble free metal ion. The sulphides of most metals with the exception of chromium are insoluble and they are readily precipitated. The carbonate salts are also only sparingly soluble at neutral pH. Toxicity by metals in industrial digesters is unlikely.

(b) *Disinfectants and detergents.* Most standard industrial detergents and disinfectants have caused problems usually as a result of a spillage or excessive use. The chlorine-based detergents such as hypochlorite and chlorine dioxide although extremely toxic at low concentrations (i.e. <1 mg/litre) (Mosey, 1982) are rapidly adsorbed onto solids and inactivated. Most common synthetic detergents are also quite rapidly broken down and concentrations greater than 20 mg/litre are required before any deterioration in performance occurs. Some detergents such as the quaternary ammonium compounds are used in the food industry because of their persistence. These can be toxic at low concentrations, about 1 mg/litre (Sahm 1984; Stronach *et al.*, 1986).

(c) *Solvents.* Most common organic solvents, alcohols, esters and ketones do not cause problems. The chlorinated solvents such as chloroform, carbon tetrachloride and chlorinated derivatives such as chloroacetic acids are toxic at 1 mg/litre (Salkinoja-Salonen *et al.*, 1983*a*). They are not easily counteracted by other reactants in the digester and should be excluded.

(d) *Biocides and pesticides.* The significantly different biochemistry of the methanogens means that they are resistant to many of the common antibiotics. Anaerobic treatment is however sensitive to agents which affect protein and lipid biosynthesis and those which interfere with membrane function such as:

Chloramphenicol	Virginiamycin
Lasalocid	Arsanilic acid
Monensin	Olaquindox
Spiramycin	Metronidiazole
Tylosin	Formaldehyde

There has been active research on the pesticides used by the forestry industry (Salkinoja-Salonen et al., 1983b) and there has been success in breaking down aldrin, lindane, dieldrin and a number of chlorophenolic compounds arising from bleaching pulp. Mixed cultures can adapt to potential toxins and recalcitrants, but inhibition is still likely from sudden changes in the concentrations of pesticides and biocides.

(e) *Ammonia.* Ammonia toxicity to anaerobic fermentation is widely reported but it is rarely a problem with industrial wastes. Farm wastes commonly contain high concentrations of ammonia, 1–5 g/litre, and most of the difficulties are with this type of waste (see Chapter 3). It is the free ammonia not the ammonium ion which causes the toxicity and the effects are therefore complicated by other reactions. The toxicity due to high shock loads of ammonia, e.g. 25 g/litre is reversible (Sixt and Sahm, 1987).

(f) *Sulphur compounds.* Sulphide toxicity of the anaerobic treatment of industrial wastes is a common problem. A wide range of fermentation processes uses molasses as a substrate. In processing molasses is acidified and pH adjusted with sulphuric acid, average sulphate concentrations are as a consequence 4–5 g/litre. Margarine and oil waste as well as pulp and paper mill effluent also have high concentrations of sulphates. There are three separate mechanisms of inhibition:

(i) One is the precipitation of essential trace nutrients (Fe, Ni, Co, Mo) by the sulphide ion. Iron chloride was deliberately added to a UK anaerobic digester treating a molasses-based waste to try and overcome the sulphate toxicity (Wheatley, 1984).
(ii) Interference of intracellular microbial metabolism by binding with essential enzymes and coenzymes. The toxic component is the unionised soluble sulphide usually H_2S and therefore its toxic concentration is a function of pH and other complexing reactions.
(iii) Competition between the methanogenic and sulphate-reducing bacteria for similar simple substrates, i.e. acetate, formate and hydrogen. Sulphate-reducing metabolism is energetically more favourable than carbon dioxide reduction to methane and the sulphate-reducing bacteria can grow more quickly than the methanogens.

Toxicity starts to be expressed when the concentration of sulphates is greater than 1·0 g/litre although acclimatisation can mean that total inhibition does not occur until concentrations exceed 4·5 g/litre. Hydrogen sulphide is released into the biogas and this does give rise to a control parameter. Concentrations of between 0·5–1·0% H_2S in the biogas are the maximum tolerable concentrations, and significant odour and corrosion problems will occur when the gas is used at these concentrations. Concentrations of up to 3–4% H_2S in the biogas have been reported (Frostell, 1981). The strengths of fermentation, margarine and pulp wastes mean that there is a strong incentive to use anaerobic treatment despite the toxicity from the sulphates. This has led to a study of two-stage systems. The first stage with sulphide generation is followed by gas stripping or sulphide precipitation before a second stage for methane generation (Anderson et al., 1986). Specific inhibition of the sulphate-reducing bacteria has also been tried. Neither of these techniques has been attempted at a demonstration scale yet.

5.9 Conclusions

Until quite recently anaerobic waste treatment was confined to complex slurries and sludges (described in Chapters 3 and 4). The recalcitrance of particulate organic matter especially lignocelluloses meant that this type of anaerobic digestion took place in stirred tanks with very long residence times. Treatment rates and methane production were too low to make the process attractive as a source of energy or more generally applicable to industrial waste water treatment. The rising cost of the energy intensive aerobic alternatives led to a surge of interest in anaerobic waste treatment both in the USA and Europe in 1975–85. There are now a variety of innovative digester designs and a much better understanding of the microbiology and control of anaerobic fermentation. In the case of some effluents, that is strong wastes from the fermentation industry, there are no economic alternatives. In other cases, i.e. dairy waste, brewery effluent, starch and confectionery, the economics are more finely balanced. The choice will depend on current practice, changes in water authority charges and local site conditions.

Some technical risks such as reliability and disposal of the residues remain to be overcome. There is also a lack of UK experience at full-scale. Figures published by the Department of Trade in the UK show spending on advancing anaerobic digestion has been about £10 million, £2–3 million has been spent on basic research and the remainder on supporting new emerging companies. There are possibilities of further government, and European Commission incentives to advance the technology particularly to overcome other environmental problems such as the greenhouse effect and waste disposal.

Anaerobic digestion processes 500 million tonnes of crude wet waste in the countries of the EC each year and it is certain that research will continue to perfect and adapt the process. Fundamental work is still necessary on the ecology and interrelationships between the groups of bacteria in anaerobic digestion. Research on the genetics and transfer of characteristics between organisms will bring long term benefits. Examples are the degradation of recalcitrant compounds and low temperature digestion. Further process engineering improvements can be expected from multistage treatment with separate hydrolysis of solids and aerobic post-treatments. Small additions of specific micronutrients will improve the process, especially aspects like granulation.

5.10 References

Alibhai, K. R. K. and Forster, C. F. (1986). An examination of the Granulation Process in UASB reactors. *Environ. Tech. Letters*, **7**(4), 193–200.

Anderson, G. K., Sanderson, J. A., Saw, C. B. and Donnelly, T. (1988). Fate of COD in an anaerobic system treating high sulphate bearing wastewater. *Biotechnology for Degradation of Toxic Chemicals in Hazardous Wastes*, ed. R. J. Scholze, E. D. Smith, J. T. Brady, Y. C. Wu and J. V. Basilico. Noyes Data Corporation, New Jersey, USA, pp. 504–31.

Anon. (1987a). Anaerobic effluent treatment plant in operation. *Water and Waste Treatment J.*, **30**(1), 12.

Anon. (1987b). Systematic sewage treatment. *Water and Waste Treatment J.*, **30**(12), 30.

Anon. (1988). Water charges: industry faces further steep rise. *Water and Waste Treatment J.*, **31**(2), 4.

BABA (1982). *Safety Code of Practice for Anaerobic Digestion.* BABA, Reading, 23 pp.

Battersby, N. S. (1989). A survey of the anaerobic biodegradation potential of organic chemicals in digesting sludge. *Appl. Env. Microbiol*, **55**, 433–9.

Bolle, W. L., van Breugel, J., van Eyberg, G. C., Kossen, N. W. F. and Zoetemeyer, R. J. (1986). Modelling the liquid flow in Upflow Anaerobic Sludge Blanket Reactors. *Biotech. Bioengineering*, **28**, 1615–20.

British Paper and Board Industry Federation (1987). *Paper and Board Industry Facts 1986.* British Paper and Board Industry Federation, London.

Butcher, G. J. (1988). Experiences with anaerobic digestion of wheat starch processing waste. *International Biodeterioration*, **25**(2), 71–7.
Butler, G. A. (1984). Anaerobic digestion of starch process effluents at Tenstar products. In *Anaerobic Treatment of Industrial Waste*, ed. J. Coombs. BABA, Reading.
Camilleri, C. (1987). Operating results from a fixed film anaerobic digester for pollution abatement and methane production from industrial wastes. In *Biomass for Energy and Industry, 4th EC Conference*, ed. G. Grassi, B. Delmon, J. F. Molle and H. Zibetta. Elsevier Applied Science, London, pp. 1338–42.
Camilleri, C. (1988). Anaerobic digestion of food processing wastewater: industrial performance of fixed film technologies for methane recovery and pollution abatement. In *Fifth International Symposium on Anaerobic Digestion: Poster Papers*, ed. A. Tilche and A. Rozzi. Monduzzi, Bologna, pp. 473–6.
Chemical Industries Association (1986). *Priority areas for Chemical Research and Development*. Chemical Industries Association, London.
Cheung, M. Y., Oakley, D. L. and Forster, C. F. (1986). An examination of anaerobic upflow filters operated in cascade sequence. *Env. Tech. Letters*, **7**, 383–90.
Dahab, M. F. and Young, J. C. (1982). Retention and distribution of biological solids in fixed-bed anaerobic filters. *Proc. 1st Int. Conf. Fixed Film Biol. Processes*. Kings Island, Ohio, April 1982, pp. 1337–51.
Demuynck, M., Nyns, E. J. and Palz, W. (1984). *Biogas Plants in Europe: A Practical Handbook*. Reidel Publishers, Dordrecht, Boston and Lancaster.
Department of Scientific and Industrial Research (1960). *Water Pollution Research, The Report of the Water Pollution Research Laboratory, 1955–1959*. HMSO, London.
Ehlinger, F., Audic, J. M., Vernier, D. and Fraup, G. M. (1987). The influence of the carbon source on microbiological clogging in an anaerobic filter. *Wat. Sci. Technology*, **19**, 261–73.
Environment Canada (1986). *Anaerobic Treatment of Dairy Effluent*. Report EPS 3/FP/1 Environment Canada, Ottawa.
Frostell, B. (1981). Applications of biological treatment methods to industrial effluents. Pilot-scale anaerobic–aerobic treatment of distillery waste. *Chemistry and Industry*, **13**, 465–9.
Gerletti, M. (1984). Anaerobic digestion of municipal solid wastes, improved by organic wastes in a large pilot plant. In *Anaerobic Digestion and Carbohydrate Hydrolysis*, ed. G. L. Ferrero, M. P. Ferranti and H. Naveau. Elsevier Applied Science, London, pp. 446–9.
Heijnen, J. J., Enger, W. A., Mulder, A., Lourens, P. A., Keijzers, A. A. and Hoeks, F. W. J. (1985). Anwendung der anaeroben Wirbelschichttechnik in der biologischen Abwasseinigung. *Wasser/Abwasser*, **126**(2), 115–30.
Jeris, J. S. (1983). Industrial wastewater treatment using anaerobic fluidised bed reactors. *Water Science and Technology*, **15**, 167–87.
Lettinga, G. and Hulshoff-Pol, L. W. (1986). Advanced reactor design, operation and economy. *Water Science and Technology*, **18**(12), 41–53.

Lettinga, G., Hulshoff-Pol, L. W., Wiegant, W., de Zeeuw, W., Hobma, S. W., Grin, P., Roersma, R., Sayed, S. and van Velsen, A. F. M. (1983). Upflow sludge blanket process. In *Proceedings of the 3rd International Symposium on Anaerobic Digestion*, Boston, pp. 139–58.

Lettinga, G., Man, A. D., de Grin, P. and Hulshoff-Pol, L. W. (1987). Anaerobic waste water treatment as an appropriate technology for developing countries. *Tribune Cebedeau*, **40**(519), 21–32.

Levenspiel, O. (1972). *Chemical Reaction Engineering*. Wiley, Chichester, pp. 253–325.

Micklewright, A. T. (1986). A review of the practice of trade effluent control and charging in the North West. *Water Pollution Control*, **85**(3), 324–36.

Ministry of Agriculture Fisheries and Food (1986). *Food Processing Research Consultative Committee. Report to the Priorities Board*. HMSO, London.

Morgan, H. (1980). The development of an anaerobic process for the treatment of a wheat starch factory effluent. In *Food Industry Wastes: Disposal and Recovery*, ed. A. Herzka and R. G. Booth. Elsevier, London, pp. 92–108.

Mosey, F. E. (1976). Assessment of the maximum concentration of heavy metals in crude sewage which will not inhibit the anaerobic digestion of sludge. *Wat. Pollut. Contr.*, **75**(1), 10–20.

Mosey, F. E. (1982). Anaerobic filtration: A biological treatment process for warm industrial effluents. *Wat. Pollut. Contr.*, **81**(4), 540–52.

Naghdy, G. and Helliwell, P. (1987). Computer aided load smoothing with energy management (CALSEM) for activated sludge treatment plants. *J. Wat. and Env. Management*, **1**(3), 339–47.

Oleszkiewicz, J. A. and Thadari, V. J. (1988). Effects of biofilter media on performance of anaerobic hybrid reactors. *Env. Tech. Letters*, **9**(2), 89–100.

Oleszkiewicz, J. A., Hall, E. R. and Oziemblo, J. Z. (1986). Performance of laboratory scale anaerobic hybrid reactors with varying depths of media. *Env. Tech. Letters*, **7**, 445–52.

Parkin, G. F. and Owen, W. F. (1986). Fundamentals of anaerobic digestion of waste water sludges. *J. Environmental Engineering*, **112**(5), 867–920.

Pearson, J. (1988). Lower costs encourage mill interest in anaerobic treatment. *Pulp and Paper International*, **3**, 28–30.

Rudd, T., Hicks, S. J. and Lester, J. N. (1985). Comparison of the treatment of synthetic meat waste by mesophilic and thermophilic anaerobic fluidised bed reactor. *Env. Tech. Letters*, **6**(5), 209–24.

Sahm, H. (1984). Anaerobic wastewater treatment. *Advances in Biochemical Engineering/Biotechnology*, **29**, 81–115.

Salkinoja-Salonen, M., Haukulinen, R., Valo, R. and Apajalahati, J. (1983*a*). Biodegradation of recalcitrant organochloride compounds in fixed film reactors. *Wat. Science Technology*, **15**(9/10), 149–59.

Salkinoja-Salonen, M., Haukulinen, R., Silakoski, L., Valo, R. and Apajalahati, J. (1983*b*). Treatment of pulp and paper industry waste waters in an anaerobic reactor: from bench to full scale operation. *Proceedings of 3rd International Symposium on Anaerobic Digestion*, Boston, pp. 17–121.

Shieh, W. K. and Mulcahy, T. (1983). Fluidized bed biofilm reactor kinetics—a rational design and optimization approach. *Wat. Science and Technology*, **15**, 197–208.

Shore, M., Broughton, N. W. and Bumstead, N. (1984). Anaerobic treatment of waste waters in the sugar beet industry. *Wat. Poll. Contr.*, **83**(4), 499–506.

Sixt, H. and Sahm, H. (1987). Biomethanation. In *The Biotechnology of Waste Treatment and Exploitation*, ed. J. M. Sidwick and R. S. Holdom. Ellis Horwood, Chichester, pp. 149–72.

Smith, M. O., Ferrall, J., Smith, A. J. T., Winstanley, C. I. and Wheatley, A. D. (1988). The economics of effluent treatment: a case study at Bovril. *International Biodeterioration*, **25**(2), 97–105.

Smith, N. (1984). Anaerobic effluent treatment in British Sugar. In *Anaerobic Treatment of Industrial Waste*, ed. J. Coombs. BABA, Reading, pp. 27–36.

Speece, R. E. (1983). Anaerobic biotechnology for industrial wastewater treatment. *Environmental Science and Technology*, **17**(9), 416A–427A.

Speece, R. E. (1988). A survey of municipal anaerobic sludge digesters and diagnostic activity assays. *Wat. Research*, **22**(3), 365–72.

Stronach, S. M., Rudd, T. and Lester, J. N. (1986). *Anaerobic Digestion Process in Industrial Waste Treatment.* Springer-Verlag, Berlin, 184pp.

Swanwick, J. D., Shurben, D. G. and Jackson, S. (1969). A survey of the performance of sewage sludge digestion in Great Britain. *Wat. Poll. Contr.*, **68**(6), 639–61.

Switzenbaum, M. (1983). A comparison of the anaerobic filter and anaerobic expanded/fluidised bed process. *Water Science and Technology*, **15**, 399–413.

Szendry, L. M. (1986). The Bacardi Corporation digestion process for stabilising rum distillery waste and producing methane. In *Energy from Biomass and Wastes VII.* The Chicago Gas Institute, pp. 767–94.

Szendry, L. M. (1983). Scale up and operation of the Bacardi Corporation anaerobic filter. In *Proceedings of the 3rd International Symposium on Anaerobic Digestion*, Boston, pp. 365–77.

Walker, I. and Austin, E. P. (1981). The use of plastic, porous biomass supports in a pseudo fluidised bed for effluent treatment. In *Biological Fluidised Bed Treatment of Water and Wastewater*, ed. P. F. Cooper and B. Atkinson, Ellis Horwood. Chichester, pp. 272–84.

Wheatley, A. D. (1984). Anaerobic digestion of waste waters, an overview of the technology. In *Anaerobic Treatment of Industrial Waste*, ed. J. Coombs. BABA, Reading, pp. 3–17.

Wheatley, A. D. and Cassell, L. (1985). Effluent treatment by anaerobic biofiltration. *Wat. Pollut. Contr.*, **84**(1), 10–20.

Wheatley, A. D., Cassell, L. and Winstanley, C. I. (1984). Energy recovery and effluent treatment of strong industrial waste by anaerobic biofiltration. In *Anaerobic Digestion and Carbohydrate Hydrolysis of Waste*, ed. G. L. Ferrero, M. P. Ferranti and H. Naveau. Elsevier Applied Science, London, pp. 284–306.

Wheatley, A. D., Johnson, K. A. and Winstanley, C. I. (1988). The reliability of anaerobic digestion for the treatment of food processing effluents. In

Anaerobic Digestion 1988, ed. E. R. Hall and P. N. Hobson. Pergamon, Oxford, pp. 135–46.

Witt, E. R., Humphrey, W. J. and Roberts, T. E. (1979). Full scale anaerobic filter treats high strength wastes. *Proceedings 34th Purdue Industr. Waste Treatment Conf.*, pp. 229–34.

Young, J. C. and McCarty, P. L. (1969). The anaerobic filter for waste treatment. *J. Wat. Poll. Contr. Fed.*, **41**(5), R160–R173.

Index

Aceticlastic, 53–7
  methanogenesis, 56, 60, 68
Acetogenesis, 51
Acetogenic, 46
Acid, 70
Acidogenesis, 208
Acidogenic, 47
  bacteria, 46
Activity, 211
Aerobic
  ditch, 103
  treatment, 95, 207
Algae, 2
Alkalinity, 70, 164, 189, 210
Alkanes, 165
Aluminium, 200
Amino acids, 58
Ammonia, 120–2, 121–2, 129, 209, 212, 217
Anaerobic filters, 10, 68, 114, 117, 178, 193, 195
Ancillary equipment, 125
Animal feeds, 184
Antibiotics, 120, 122, 166
Antibodies, 70
Aromatic, 7, 52
Ascaris, 9
  see also Roundworm
Aspect ratio, 161–2
Auger, 119
Autotrophic, 62
Avoiding waste, 184
Axenic, 47

Bacterial retention, 191

Baffles, 202
Baking, 151
Batch digesters, 105
Bearings, 150
Bicarbonates, 5, 7
Biocides, 122
  pesticides, and, 166, 216
Biodegradability, 182
Biofilms, 60, 68
Biological tests, 210
Biomass
  concentration, 9
  measurement of, 70
  substrate, as, 9
Blockage and channelling, 196
BOD, 208
Boiler, 126
Brewing, 179
Buffer, 165, 210
By-product, 9, 96, 179, 184

Calcium, 200
Calorific value, 5, 6
Calorimetry, 163
Capital costs, 31, 213
Carbohydrates, 100, 102
Carbon dioxide, 70
Carbon monoxide, 54
Carboxylic acids, 2, 70, 72, 103, 121, 164
Cardboard, 102
Celanese, 182
Cellulolysis, 116
Cellulose, 50, 101

Charging formula, 175
Chemical analysis, 183
Chemoheterotrophic, 12
China, 104
Chlorinated aliphatic, 52
Chlorinated aromatics, 52
Chlorinated solvents, 166, 182
Clogging, 196-8, 203
COD, 208
Coenzyme, 211
  Coenzyme F_{420}, 71
  Coenzyme M, 57
Compost, 106, 119
  process thermophilic, 106
Concentric tube, 151
Condensation, 7
Conductivity, 209
Consent, 188
Construction, 16, 124
Constructors, 5
Contact digesters, 113-14
Contact process, 10, 68, 178
Contact stirred tank reactor, 191, 193
Contract type, 12
Control, 56, 189
  monitoring, and, 163
Corrosion, 16, 22
  scales, and, 152

Dairy, 103
  industry, 177
Damage, 16
Danger, 103
  see also Safety
Design, 16, 185, 189
  manual, 145
Detergents, 166, 182, 191, 216
Different, 34
Digester
  crusts, 118
  surveys, 145
Digestion, 22
Dilution, 16
Disinfectants, 120, 123, 143, 181, 216
Dispersion, 149
Distillery, 179
Distribution, 200

Distributors, 198
Downflow filters, 196
Drinking water, 129

Economics, 10, 14, 20, 22, 183, 195, 211
  analysis, 12
  industrial digesters, 26
  landfill gas, 31-2
  municipal solid waste, 30
  pay back, 12
  rural digesters, 14
Effective biomass, 165
Energy, 96
  resources, 3
Engines, 126-7
Enhancement, 33
Esters, 7
Ether, 165
European design, 146
European legislation, 167
European practice, 140
Expanded bed, 68
Explosive gas, 95
  see also Safety
Extremeophiles, 10

Falling film coolers, 162
Farm digesters, 15, 18
  examples, 110
  surveys, 18
Fatalities, 104
Fats, 100, 102, 191, 196, 199
  oils, and, 190
Fatty acids, 72
Fed-batch, 107
Feeding, 128
Fertilizers, 7, 96, 184
Fibre, 100, 115-17
Filter, 113
Filters. See Anaerobic filters
Flame traps, 156, 208
Floating solids, 118
Flocculated granulated biomass, 197
Flocculated-biomass, 114
Flocculation, 178

Flocculator, 195
Flocs, 60
Flotation, 190
Flow balancing/two-stage treatment, 188
Fluidised, 113
 bed, 68, 114, 203
 or expanded bed, 193
Foam, 7, 203
Formate, 48, 60
Fouling, 151
Fruit wastes, 114

Gas
 composition, 126
 flares, 156
 recirculation, 148
 sales, 30
 storage, 125, 155, 206
 use, 33
 yield, 102, 209
Granulated, 202
Granulation, 199, 200
Granules, 60, 68, 203
Grasses, 106
Grit, 22, 118–19, 149, 188
Growth rates, 112

Halogen, 7
Hay, 106
Heat
 exchangers, 109, 150, 206
 losses, 152
 transfer, 151
Heating, 16, 127, 150
Hemicellulose, 101
History, 143
Homoacetogenic, 46
Homoacetogens, 46–7
Horizontal digesters, 111
Hybrid, 115, 193
Hydrocyclones, 188
Hydrogen, 7, 56, 72
Hydrogen sulphide, 208
Hydrolysis, 50, 178
 acidogenesis, and, 45
Hydrolytic bacteria, 46

Impeller mixers, 148
Inclined plate separators, 191, 193
India standard design, 104
Industrial effluent, 24
Infra-red spectroscopy, 163
Inhibition, 214
Inhibitors, 165
Instrumentation, 208
Insulation, 124–5, 152, 206
Intensive farming, 94
Intermediate alkalinity, 165

K_s, 53

Labour, 213
Lagging, 152
Lagoons, 193
Land irrigation, 179
Legislation, 11, 142, 173
Lignin, 2, 165, 182
Ligno-cellulose, 51
Loading municipal solid waste, 29
Loading rates, 7, 26, 29, 145
 CSTR, 194
 filters, 198
 fluidised bed reactor, 204
 industrial digesters, 26
 VASB, 202
Low temperature digestion, 107

Malodorous, 95, 121
Market size, 4–5
Materials of construction, 204
Mathematical models, 59, 130
Meat processing, 181
Membrane filtration, 193, 195
Mercaptans, 7
Mesophilic, 46–7, 62, 150
Metal
 requirements, 65
 toxicity, 166, 216
Methane, 5, 126, 212
Methanogenesis, 51–2, 54, 208
Methanogenic, 52
Methanogens, 48

Microalgae, 102
Microbial ecology, 200
Milk waste, 114
  see also Dairy
Minimum size, 104
Mixing, 16, 109, 128, 194
Monitoring, 56, 70
Monoxenic, 47
Municipal solid waste, 2, 28

Newspaper, digestion of, 102
Nitrogen
  fertiliser value, 7, 124
  fixation, 63
  requirements, 63, 101
Nutrient requirements, 62, 71, 185, 190, 214

Odours, 9, 11, 27, 31, 94–5, 150, 207
Oils, 191
Organic loading rate, 9, 195
Oxygen, 7

Packing material, 196
Paper industry, 181
Partial alkalinity, 165
Pathogens, 9, 96, 142
Payback, 12, 26, 32
Performance indicators, 11
Pesticides, 182
Pharmaceutical wastes, 183
Phosphate requirements, 64, 71
Phosphorus in granulation, 200
pH, 70, 164, 189, 210
Picket fence stirrer, 153
Pilot studies, 185
Plant life, 156
Plants, 173
Plastics, 161
Plug-flow digester, 110
Potatoes, 102
Power requirements, 148, 150
Prefabrication, 159
Preflocculation/degassing, 193
Pressure, 156
Problems, 16–17, 22
  See also Reliability

Process risk, 161
Procurement, 166
Protein, 2, 102, 129
Protozoa, 60
Pumps
  feed type, 119

Rag, 149
Rashig rings, 196
Recalcitrants, 182
Recycling, 184
Reliability, 67, 198
Relief valves, 208
Renewable energy, 2
Research
  development, and, 4
  landfill gas production, 33
Residues, 128
Retained biomass, 103, 111
Retention times, 7, 108, 112–13, 116–17, 119, 143, 157, 190–1, 198
Roundworm, 9
Running costs, 173
Run-off, 94
Rural digesters, 12

Safety, 95, 103–4, 123, 207–8
Safety valve, 128
Sampling, 183
Sawdust
  digestion of, 102
Scraping mechanism, 190
Screening, 188
Screw feed, 119
Sealing paints, 125
Seasonal processes, 178, 180
Seaweeds, 102
Sensors, 162
Separation, 186
Settlement, 190
Settling velocities, 199
Sewage
  sludge, 21–2
  treatment, 140
  works, 22, 180

Shock loads, 2, 182, 197, 214
Silage, 102, 213–14
Silicon in granulation, 200
Sludge
  blanket digesters, 113
  consolidation, 153
  feeding, 155
  production, 212
  recirculation, 148
Slurry solids, 102
Smell. *See* Odour
Solids concentration, 119
Solvents, 216
Species of methanogens, 4
Specific growth rate, 53
Spiral tube, 151
Starch, 178
Start-up, 123, 182
Steam injection, 152
Stirred tank, 108
Submerged combustion, 152
Subsidies, 11
Substrate affinity, 53
Sugar confectionery, 178
Sulphides, 212
  toxicity, 61, 62, 126, 180
Sulphur requirements, 65, 101, 217
Survey industrial digesters, 24, 27
Surveys
  biogas, use of, 7
  digesters, of, 143, 157, 194
  energy crops, 34
  farm digesters, 16
  landfill gas, 33
  sewage sludge, of, 22
Symbiosis, 198

Tapeworm, 9
Tax credits, 11
Temperature, 120

Terpenes, 6
Thermal conductivity, 163
Thermophilic, 47, 52, 54, 62, 150
  digester, 121
TOC, 208
Toxicity, 122, 165, 207
Toxins, 185, 214
Trace metals, 215
  See also Metals
Training, 185, 213
Treatability, 179–81, 185
Trickling filter, 103
Tubular digester, 110
Turbidity, 209
Turn key contract, 12, 161
Turnover, 149
Two-phase, 113, 115
Two-stage
  cultures, 115
  treatment, 72, 115, 178

Uncoupling, 65, 71–2
Upflow anaerobic sludge blanket (USAB), 10, 68, 178, 193, 199
Upward flow velocity, 202

Vacuum, 156
Vegetable wastes, 100, 114, 180
Vehicle fuel, 3
Vertical tube digester, 108
Vitamins, 58, 62
Volatile solids, 163
Volume of farm waste, 97

Waste audit, 183
Water
  usage, 144, 176
  weeds, 2